Veritocracy

Collective Studies in Knowledge and Society

Series Editor

James H. Collier, Virginia Tech

This is an interdisciplinary series published in collaboration with the Social Epistemology Review and Reply Collective. It addresses questions arising from understanding knowledge as constituted by, and constitutive of, existing, dynamic and governable social relations.

Recent titles in the series

Veritocracy: Truth, Science, and How to Preserve Democracy, by Harry Collins and Robert Evans

Philosophical Interventions in Neoliberal Higher Education, edited by Brandon Absher

Interdisciplinarity and the Art of Human Connection: Knowing in a Complex World, by Sharon Woodill

Institutional Epistemology and Extreme Inequality: Knowledge and Governance in a Non-ideal World, by Marko-Luka Zubčić

Rorty, Public Reason, and Modernity's Crisis of Critique, by Ivan Marquez

Disagreeing Despite the Data: The Destruction of the Factual Commons, by David Apgar

The Epistemology of Disasters and Social Change: Pandemics, Protests, and Possibilities, by Jordan Pascoe and Mitch Stripling

For and Against Scientism: Science, Methodology, and the Future of Philosophy, edited by Moti Mizrahi

The Social Production of Knowledge in a Neoliberal Age: Debating the Challenges Facing Higher Education, edited by Justin Cruickshank and Ross Abbinnett

Democratization and Struggles Against Injustice: A Pragmatist Approach to the Epistemic Practices of Social Movements, by Justo Serrano Zamora

Minority Report: Dissent and Diversity in Science, by William T. Lynch

Epistemic Paternalism: Conceptions, Justifications and Implications, edited by Guy Axtell and Amiel Bernal

Veritocracy

Truth, Science, and How to Preserve Democracy

By Harry Collins and Robert Evans

BLOOMSBURY ACADEMIC
NEW YORK • LONDON • OXFORD • NEW DELHI • SYDNEY

BLOOMSBURY ACADEMIC
Bloomsbury Publishing Inc, 1359 Broadway, New York, NY 10018, USA
Bloomsbury Publishing Plc, 50 Bedford Square, London, WC1B 3DP, UK
Bloomsbury Publishing Ireland, 29 Earlsfort Terrace, Dublin 2, D02 AY28, Ireland

First published in the United States of America 2026

A catalog record for this book is available from the Library of Congress.

ISBN: HB: 978-1-6669-8090-5
ePDF: 979-8-216-26602-0
eBook: 978-1-9787-5965-7

Series: Collective Studies in Knowledge and Society

Typeset by Integra Software Services Pvt. Ltd.

For product safety-related questions contact productsafety@bloomsbury.com.

For Cat Squad: Collins's grandchildren

Contents

List of Figures

List of Tables

Acknowledgements

For insights and help we want to thank Naomi Oreskes, Bryn Jones, Phil Mirowski, Deirdre McCloskey, Luis Reyes-Galindo (especially in respect of Ch 5), David Ludwig, Bryan S. Turner, Tony Mills, Wiebe Bijker, Will Mason-Wilkes, Aaron Thierry and Josh Ryan-Collins. Some of the insights came in a more roundabout way than others, some were not altogether willingly volunteered and all misunderstandings, mistakes and infelicities are the responsibility of the authors. Harry Collins also thanks Susan Collins for originally coming up with the term 'veritocracy' which suddenly pulled together an amorphous set of thoughts. Evans is grateful for the invitation, funded by the Deutsche Forschungsgemeinschaft (DFG, German Research Foundation) – Project 470816212/KFG43, to spend the autumn of 2023 as a visiting fellow at the SOCRATES centre at the University of Hannover.

Introduction

This book is about the West's potential slide into fascism and how science might help to stop it. The slide might be well advanced by the time this book is published, the 2024 electoral success of Donald Trump, with his cabinet choices and Project 2025 hovering in the wings, being a precursor of what has happened, but there may still be chances to reverse it in the long term and bring back democracy. The universal role of truth will not change whatever happens to it in today's world; in terms of politics, its role is still the same as described by Hannah Arendt in respect of the Fascist and Communist totalitarianism of the middle period of the twentieth century. '*The ideal subject of totalitarian rule is not the convinced Nazi or the convinced Communist, but people for whom the distinction between fact and fiction (i.e., the reality of experience) and the distinction between true and false (i.e., the standards of thought) no longer exist.*' ***Hannah Arendt*** (1951) ***The Origins of Totalitarianism*** *(p. 474)*

A lot of books discussing the demise of Western democracy and possible remedies have been published – at least fourteen and probably many more since 2017 – but this book concentrates on the culture of democratic society and is the only one that focuses on the role of science in that culture. America is an iconic example of the changes we are looking at. According to the recent, 2023, book by Steven Levitsky and Daniel Ziblatt, *The Tyranny of the Minority*, America has fallen behind as a democracy when compared to other Western states because its Constitution was a compromise intended to mollify the minority of slave-owning states and its rules of procedure seem designed to stifle modification and modernization; almost every attempt at reform has failed. Nowadays these same rules are being crudely exploited by the Republican Party. We argue that this is because of what we call 'The Law of Conservation of Democracy', which says *you cannot get more democracy out of a nation than you put in*. If not enough understanding of democracy has been put into the society, then undemocratic actions by politicians such as lying and law-breaking will not be met by a sufficiently massive reaction amongst the public. Levitsky and Ziblatt are surely right about the procedural changes that are necessary if American democracy is to be saved, just as Peter Turchin's book, published in 2023, is surely right about the material change that will be needed – modifying the 'wealth pump' and moving us back towards the 'post-war settlement' – but we see these changes as something that are going to be possible only if the cultural context will allow them. And that means culture is going to have to change if democracy is going to be rediscovered, secured and maintained. Here we are looking at ways to put more democracy into the culture of society, with a focus on how science can counter the erosion of truth.[1]

We claim here that a properly understood science can support a democracy based on truth – a 'veritocracy'. There is a huge and growing literature on the relationship between science and society, but our focus is on the most general question: science as an institution that upholds truth as an object lesson for politics and the wider society. Science, if it is properly understood and executed, works in two ways in its interactions with democracy: if its substantive findings are accepted, then, with the effectiveness of disease prevention and climate change as current examples, it acts as a check and balance on the technological proclamations of dictators and conspiracy theorists. But still more fundamentally, as an institution whose raison d'etre is the search for truth, it exemplifies the place of truth in all democratic decision-making.

What we are advocating in respect of politics is orthogonal to the usual left-right tension: we are pressing for a politics founded on truth, believing that a veritocracy – a democracy with truth at its heart – is incompatible with either left or right totalitarianism.

To make the case we must cover the nature of science, the nature of democratic politics, the way societies work and how all these things combine. We will have to appeal to a range of audiences if the analysis is to have the impact it needs. The audiences include fellow academics in sociology, history, philosophy and political science, especially those who have come to believe that science should no longer be treated as a trustworthy institution but is itself a form of politics to be judged by the people; we have to persuade them that there is a way out of the problems of postmodernism on the one hand, and a way out of identity politics on the other, that does not require us to sacrifice too much of what has been learned over the last half-century about the uncertainties of science. We know that to go anywhere this must strike a chord, either directly or indirectly, with influential actors such as print and broadcast journalists and commentators, political parties, such businesspeople as realize that the future we want depends on the survival of democracy and, crucially, educators – we'll suggest some changes to civic education. This will have to lead on to a change in citizens' home environments with a better understanding and appreciation of science in both younger and older generations, perhaps triggered by deeper consideration of vaccination and global warming. We hope all this will eventually feed into what counts as attractive political campaigning, with truth replacing sloganizing and, eventually, with the growing role of truth in society as a whole, replacing the catastrophic erosion of truth we see today.

What we are advocating in respect of science goes in the opposite direction to various contemporary trends which try to integrate public values into science; in contrast, we think the values of science need to be integrated into public life. The trouble is that since the 1960s science can no longer be thought of as an infallible producer of truth even if that remains its foundational aspiration. Nevertheless, it is the aspiration that makes science the way to bet even when success is not guaranteed: where truth is the goal, it is wiser to bet on those for whom it is an aspiration than on those for whom it is not an ultimate goal. Worse, the label 'science' or 'scientific' can cover a range of activities found in contemporary society, some of which do not have truth as an aspiration – these are not really science at all. We'll work our way through different classifications and types of science with the essential feature of the institution in mind.

This book is mainly about what is happening now in rich countries leaving discussion of how poor countries are exploited by science and its products to the many others who are studying it and writing about it. What we want to preserve in 'The West' is democracy and this means resisting the erosion, in political life, of the difference between the true and the false that was recognized by Hannah Arendt as fertile ground for the growth of the totalitarianism in the 1930s. Instead, we want to establish the idea of a veritocracy and nurture veritocracies.[2] The excuse for our narrow focus on Western countries and on the here and now is that if the powerful Western nations slide into populism and fascism, the closest historical precedent will be the situation just before the Second World War, except that, unlike then, *all* the powerful countries will be on the wrong side.[3] So, there will be no *Third World War*, it will have been lost before it begins, but the poor countries currently exploited by the scientific products of the rich countries will be looking back fondly to the day when there was any point or even possibility of protesting about it; our problem, if it is not solved, will be their problem too. The same goes for those whose current concern is the way that well-educated or experience-based activist citizens can interact with the scientific establishment; under fascism the only expertise will be that of the demagogue.

One inspiration for this book is a series of lectures by one of sociology's founders, Emile Durkheim, which spanned the end of the 1800s and the start of the 1900s and were subsequently published in various editions of a book entitled *Professional Ethics and Civic Morals.* Durkheim was concerned with the destabilization of society caused by industrialization and proposed that the moral integrity of the professions might act as a countervailing factor. Now that the professions are something different to what they were in Durkheim's day, our fractal model approach, described in Chapter 3, can be seen as a generalization of Durkheim's idea.

Where does the book stand in the larger discussion of the decline of democracy?

Three kinds of explanation

As mentioned, there have been at least fourteen books published recently that discuss the causes of the decline of democracy. In the final chapter we will list them (Table 7.2). There are three broad ways of explaining the decline of democracy: we'll label them the institutional, the material and the cultural. The first kind of explanation is what we are calling *institutional explanations*. These are the proximate causes of the demise of democracy because they refer to the behaviour of those within, or close to, the political system that does the governing and embodies the way governing is done. Thus, lately, political behaviour in the United States has become degraded. Among the Republicans this includes what can be described as 'working to rule'. Written constitutions, or indeed any set of formal rules, do not contain the rules for their own application and it is always possible to exploit such rules in extreme ways that were never intended by those who put them in place. Trade unions when they decide not to strike, but work to rule instead, know it will have a devastating effect, but all they are doing is sticking to the rules in pedantic ways. During the presidency of Biden, members of the US Republican Party were mostly interpreting the rules so as to exercise power and prevent the Democrats governing, rather than to smooth the path of government. Institutional explanations rest on the willingness of individuals associated with governments to change their practices in ways that serve themselves rather than democracy. The institutional forces leading in the direction of the collapse of democracy in the United States are readily visible just as they were visible in the rise of Hitler, but in terms of the three-way analysis, they are continuous with coups d'état and any other kind of violent assault on the governing institutions of a democracy.

Material explanations are the second way of explaining the decline of democracy. Material explanations are a step away from the proximate cause of change; they refer to changes in the conditions in which the citizens of a nation find themselves. For example, Marxist theory argues that the decline of democracy is inevitable as a result of the logic of capitalism leading to increasing immiseration and polarization among the population to the point where violent revolution is bound to be the outcome: of course, some group still has to lead such a change, and that is the proximate cause, but it will fail unless the population is ready to support it. A less historicist version of this kind of economic explanation is put forward by many modern analysts: since the 1970s, inequality of wealth in the West has been increasing to an extent that was quite unexpected to those familiar with what was called the 'post-war settlement', leading voters to feel disempowered and disenfranchised. Peter Turchin's (2023) cliodynamic theory is one of the most convincing of these treatments based on economic considerations and avoids the 'historicist' criticism of Marxism being based on the analysis of a hundred or more detailed historical cases. Economic strains can be exacerbated by rapid changes in the demographic structure of a nation such as the attempt to enfranchise Black citizens after the American Civil War or rapid increases in migration, also a factor in the United States. A sense of disempowerment can lead citizens to reject democracy or cease to care about it. If citizens don't value democracy, it will leave space for non-democrats to exploit.[4]

Cultural explanations are the third way to explain the change we are looking at. Cultural explanations move still further from the proximate causes of change. The envelope of legitimate political behaviour in a society is set by the culture. There are societies where large-scale corruption backed up by state-sponsored violence is an accepted part of political life, but these things are not acceptable in democracies. But culture can change. A cultural explanation would note that Western democracies are becoming more tolerant of what once would have been thought of as unacceptable political behaviour and when the citizens no longer understand what behaviour compatible with democracy means, democracy will not survive. Note that cultural change and institutional 'hardball' such as a work to rule can work hand-in-hand, exemplified by President Trump's claim that the election that he lost was stolen. He continued to insist on this in the face of a series of court cases that showed his accusations of corruption were baseless. By continuing to erode truth in this way, Trump is bringing about a cultural change that will make once unacceptable institutional change acceptable – the erosion of truth paving the way for the demise of the rule of law.[5] Likewise, material changes and cultural changes can work together because, for example, in the case of demographic change, it is the popular perception of the impact of immigration that counts: for example, in Britain immigration that supports the harvesting of farm produce, that supports the health service, that provides caring services for the elderly and that supports the economy in many other ways, is still presented as undesirable.

The fourteen books listed in Table 7.2 each tend to emphasize one of the three kinds of cause, the emphasis being divided up about equally among them. But the interaction of various causes is seen in most of them. Thus, the books by Levitsky and Ziblatt focus on the proximate institutional cause but, as their second book shows, the institutional forces that, after the American Civil War, destroyed a Southern democracy that momentarily represented the entire population could have operated only in the cultural environment of widespread and violent racism. The influence of racism, bolstered by demographic changes, now allied with evangelical Christianity, has not gone away, while whites' fear of losing the America they have grown up with underlies Trump's electioneering and the conversion of the Republican Party to one which is willing to see democracy wrecked for the sake of power. So, according to those authors, it is a cultural view concerning the relationship between white and non-white Americans which underpins the institutional lurch away from democracy. Of course, there is always continual feedback from institutional power to the national culture.

In terms of the three-way classification, this book, *Veritocracy*, has a marked and relatively narrow focus. It concentrates on the cultural environment of politics and the role of truth within it. This is because it emerges out of a concern with the social and philosophical basis of knowledge and where science stands within knowledge as a whole. It concentrates on the erosion of truth in academic and public life, and on science as a way in which the erosion might be reversed; if truth goes, civilization goes. Exploitation of the institutions of government might be the proximate causes of the decline of democracy, backed up by the effect of material changes on the citizenry, but we argue that given a sufficiently robust democratic culture, the levels of inequality we have seen since the 1970s would not have come about, the relationship between races would have been perceived differently, and there would have been no support among voters and no support among politicians for the way political institutions have been misused. And, as we will see in Chapter 7, the citizenry can express their views effectively between elections, as they did during the course of the last UK Conservative government just by indicating how they are intending to vote. If politics is downstream of culture, and in accordance with the way in which thinkers from both left and right and their media-mogul supporters act, then culture, and truth's role within it, is a crucial contributor if democracy is to be preserved in the long term. This book addresses the most distal cause of democracy's demise, the demise of truth as a component of culture, which amounts to the demise of culture itself.

The chapters

Chapter 1 lists alternative descriptions of science and their relationship to truth with democracy in mind, while the term 'veritocracy' is introduced and distinguished from similar-looking terms. Chapter 2, which will be largely familiar to professionals in the science and technology studies domain, is a brief historical sketch of what has happened to the idea of science over the last half-century, explaining how the descriptions of science as a near perfect kind of practically expressed logic, which dominated the beginning of the period, gave way to a more culturally engaged model which encouraged the melding of experts and society. That chapter ends by suggesting an alternative to this continuing slide from authority to supposed democracy in the understanding of science.

The first two chapters present what we consider to be the important features of science. Chapter 3 is about the nature of society, our description turning on 'the fractal model'. Chapter 4 introduces a new analytic classification of science aimed at showing the particular places and ways in which the general public can and can't contribute to science without it collapsing into politics. Chapter 5 introduces the idea of 'virtual diversity' and aims to strengthen science by enabling greater diversity but without violating the norms of uniformity and universality. Chapter 6 focuses on economics as an example of a type of science that has helped to give rise to a kind of activity, first introduced in Chapter 4, that we call 'ubiquitous science'; this is a so-called science which involves the public in ways that vitiate science's role as the foundation of truth in society. Economics is also troubled by the reification of mathematical models to the point where they seem to become the world rather than just an aid to understanding the world, vitiating the search for correspondence truth which is the heart of our definition of science. Chapter 6 also follows existing analyses of the development of economics in 'The West' in seeing it as the foundation of neoliberal politics with its obsession with the efficiency of markets. Under neoliberalism, market forces replace professionalism as the way experts make their choices and this, it has been suggested, is why the public has lost its respect for professional institutions and, consequently, for truth. Chapter 7 finds some hope in recent events and looks at what else must happen if the ambitions expressed in this book are to be realized, including a change in civic education concerning the role of science.

1

Truth, political culture and alternative descriptions of science

The bottom turtle

Figure 1.1 The bottom turtle (created by Harry Collins).

Those who study knowledge mostly want to understand its foundations. In the old story, illustrated in Figure 1.1, the novice asks the sage, 'what does The World stand on'; the sage replies, 'the shoulders of Atlas'. 'What does Atlas stand on', asks the novice; 'an elephant', replies the sage; and 'what does the elephant stand on', asks the novice; 'a turtle', replies the sage; 'and what does the turtle stand on', asks the novice. The sage replies, 'from there on its turtles all the way down'. Taking The World to symbolize knowledge, we want to find the bottom turtle. What is the foundational justification for what we know?

Up until half-century or so ago, many academics were pretty certain that the bottom turtle – the location of the most perfect knowledge possible – was science. If you want to understand the foundation of knowledge, science provides it through its combination of observation and experiment: replicate an experimental result and it is true, if you can't repeat it, it is not true. But since the 1960s and 1970s, science has been shown to be wanting in terms of 'bottom-turtleness'; observation and experimental findings turn out to rely on all manner of 'non-epistemic' judgements and interpretations, leaving science looking somewhat less than perfect. We'll explain some of this more fully as the chapters unfold. In particular, the revised understanding of science leaves scientific knowledge open to influence by subtle cultural prejudices and trends, with the consensus that emerges reflecting the social context.

Many contemporary academics in sociology, philosophy and humanities don't think there is a bottom turtle – attempts to justify knowledge go down and down for ever: that arguments come to an end is a matter of social agreement, not theoretical, philosophical or experimental inevitability. Nevertheless, even if this is the case, there are good ways and bad ways to reach agreement. We are arguing that telling the truth is the basis of a good way to reach agreement whether it is essentially social or not. Agreeing that truth is important is to get quite near to the bottom turtle. If we cannot find the very lowest turtle, truth is pretty near the bottom, and we can work our way back up from there.

That lowest turtle we can find, then, is not found in the domain of physical truth, but in the realm of the moral. In an earlier book we illustrated what we meant by saying that almost no one thinks that the gratuitous torturing of children is anything other than bad – as near to an absolute truth as you can get; the evilness of the gratuitous torturing of children is close to a bottom turtle, and closer to the bottom than anything discovered by science.[1] To encapsulate the argument of this book, the value of truth itself is even closer to the bottom. It is the starting point for knowledge and the foundation of language, culture and morality of all kinds.

The question then becomes how to resist the erosion of truth and truth-telling. Once it might have been the case, as Durkheim suggested, that the professions could fulfil this role but neoliberalism has replaced professional standards with market outcomes degrading the professions across the board.[2] But science, though under threat, could be an exception so long as it can maintain the centrality of curiosity about the observable world in the face of demands to become just another marketable commodity or political pawn. Science's goal – its 'formative aspiration' – as we will try to establish more carefully later in this chapter, is collectively to discover the truth about the observable world. And individual scientists soon find out that the collective goal will be elusive unless they cleave to moral truth – reporting the results of their individual work clearly and honestly. That, we will argue, is why science is a crucial institution in a veritocracy.[3]

Truth

For there to be human society of any kind, people must intend to tell the truth most of the time. If one has no idea whether someone is trying to speak truthfully, then language does not work as a means of communication and there can be no humanlike society. Truth begins in babyhood with carers' babytalk creating a consistent world of objects and sounds which evolve into language. This can, however, merely set the scene for the erosion of truth in later life with language being used as a tool of oppression by those who have control over meaning. George Orwell's 'Ministry of Truth' captures the idea.

The erosion and control of truth is not the same as lying. If you know someone is lying consistently, it is possible to extract quite a lot of meaning from what is said. In contrast, when the distinction between truth and lies disappears, there is no basis for inferring from the past into the future. Instead, there is only what is said here and now, which may or may not be what is said tomorrow: one might say that in these circumstances, the philosophical 'problem of induction' has metamorphosed from an abstract idea to a regular member of the family.[4] For those in power, language used this way is just another weapon in the armoury, defining how we are to think and act. For those on the receiving end, it is Orwell who captures it once more: resistance eventually becomes futile, and peace of mind can be found only by loving the leader and their preferred moment-to-moment reality. This is the outcome, not of lying, but of the disordering of meaning.

The idea that truth matters is under attack across the world as both new and old democracies become more authoritarian. The United States is the most concerning example of this trend because it is the most powerful nation and because it is the traditional supporter of democracy in troubled times. But there are a surprising number of smaller countries whose populations seem ready for a return of populism and fascism. In this strange world, President Nixon is an example of respect for truth. He was a determined, and for a time, successful, liar, but that was because he had enough respect for the idea of truth to want to try to simulate it by cleaving to a consistent story if not a true one; he wanted to create an orderly but false reality. In contrast, President Trump, like President Putin, uses language without any regard for consistency except to consistently serve political convenience and this is done best by disordering meaning, so it can be recreated moment to moment.

Forthright displays of disregard for logic and correspondence with reality are used by the powerful to show us where the new truth is located – it is located with the powerful, not with the way things are! This isn't just a more extreme form of what politicians have always done; it is a new way of doing politics, constrained only by how things are said to be. We say a new way of doing politics, but it is, of course, familiar to those with long memories: it is the German Nazis' 'big lie' polished up with postmodernism and beefed up with the internet. To repeat, we want to counter the erosion of truth in Western societies by changing the culture to one where truth-telling is the expectation. Given that we can't rewind history, we need to find a way that acknowledges that the widespread scepticism about contemporary democracy and the place of science within it is revealing, but we want to find the solution in making democratic institutions, including science, stronger, not weaker. Monotonic, unrelenting scepticism can only weaken.

Introducing veritocracy

Both the political left, as exemplified by Antonio Gramsci, and the political right, as exemplified by Andrew Breitbart, agree that politics is downstream of culture: the envelope of viable government is set by the culture of the society.[5] Radical political change of the kind that is emerging in the United States and elsewhere is made possible by using the media – both social and broadcast – gradually to change the discourse of the wider society and this changes what it is possible for politicians to say and do. In the final chapter we'll redescribe some very recent events, arguing that they show how the remnants of more traditional cultural values within both the UK and United States have rescued us from the worst excesses of this kind of politics. But the relief in the United States has turned out to be a momentary because the balance in the United States has become narrow to vanishing and that momentary respite has been swamped by democracy eroding forces.

Western culture needs to move in the opposite direction. Culture and language are intimately related, and the new word, 'veritocracy', symbolizes what we mean. A veritocracy is a democracy in which truth is a central political value. If the term 'veritocracy' were to gain currency and respect, it could begin to help in the battle for truth.

Of course, such words can be used inauthentically, but hypocrisy is a start on the right road; Nixon was a hypocrite; his hypocrisy was, in the 300-year-old words of La Rochefoucauld, 'the tribute vice pays to virtue'. Hypocrisy about truth isn't good but it is better than the erosion of the difference between truth and lies. If politicians were to claim more frequently that it is important to tell the truth, then they would find themselves having to do it more often and to be more accountable for it to the citizenry. It won't be one word alone that makes the difference, but the new network of discourse and debate that will develop with it. Spoken language carries moral and practical consequences. New language is new culture, and politics is downstream of culture. As a commentary by Sam Power on UK Prime Minister, Rishi Sunak's pledge to rule with integrity put it:

> The failure of now-sacked Tory party chairman Nadhim Zahawi to pay his taxes strikes at the heart of what irks the British public most about their politicians. The general sense that there's one rule for us and another for them has been spreading since the middle of Boris Johnson's tenure. That's precisely why Rishi Sunak pledged to bring integrity back to Downing Street upon taking office. But like John Major in the 1990s, he appears to have made a rod for his own back with this promise. Now he has to answer for all the bad behaviour, whether he knew about it or not.
>
> (Power 2023)

We see signs in the UK that the seeming storm of untruths under Johnson along with the embrace of ideology and slogans under Truss is already creating a reaction that is moving us in the direction of veritocracy. Keir Starmer, the leader of the opposition in the UK, has also proclaimed that the future will be characterized by integrity and the outcome of the 2022 American mid-term elections momentarily indicated that the erosion of truth is not as attractive to voters as had been feared even though the proportion it does attract it still terrifyingly large and has proved to be decisive in more recent events such as the 2024 US election – the respite is over. So now we have to look to the future for the idea of science, if properly understood and executed, to help consolidate any fragile cultural move to veritocracy.

The term 'veritocracy'

Why 'veritocracy'? Veritocracy is 'rule by the truth'. We know it is a word that is going to trigger negative reactions in social and political scientists and philosophers as they reach for their standard objections to anything that might take power away from the 'people' and give it to an elite. But veritocracy is not 'technocracy', rule by technical experts, nor is it 'epistocracy', rule by the knowledgeable. The reason is that veritocracy is not a formula for making decisions, or even for identifying who should make the decisions, but a moral prescription for the culture of decision-making. It says that truth-telling should be at the heart of politics, but it does not say what the substance of truth is. Where a technical decision has to be made, or where technical information bears on a decision in some way, then all veritocracy says is that the views of those who can be best trusted to tell the truth about the technical matters should be given special importance and politicians should tell the truth about what the truth-tellers have said to them and how they are reacting to it.

This is not the same as saying that those who provide the technical information should take the decisions. To repeat, it implies only that the political decision-makers should (a) take the truth-tellers' views into account and (b) explain, truthfully, to the citizens what those views are and how the political decisions relate to them, especially if there is disagreement with the apparent technical drift of the matter. For instance, President Trump's decision to make policy that clashes with climate change science is not an unreasonable thing for a politician to do but he should not be attacking climate change scientists or claiming that the science of climate change is fallacious or a left-wing conspiracy. Climate change science is the best science we are going to get on the matter and the best bet for being the truth of the matter. But Trump would still be politically entitled to state this and then say, 'but I don't care', we're still going to 'drill baby drill'. Then voters could consider whether they preferred safeguarding future generations to generating wealth today – which is how the science looks – and without being confused by being told that the science is wrong.

A way to see this is to understand that the term 'veritocracy' arose as a reaction to President Trump and, to a lesser degree, Prime Minister Johnson. Both of them appeared uninterested in truth to the extent that they were not even concerned with creating a systematic campaign of lies. Truth, it appeared, was simply not a feature of life in which they invested any importance except to ignore it and encourage everyone else to ignore it. The iconic moment came right at the start of Trump's regime, following the claim by his Press Secretary, Sean Spicer, that Trump's inaugural crowd was larger than had been present at Obama's inauguration. To those still cleaving to the old idea of truth, this was preposterous as it conflicted with aerial photographs and transit data. But, when challenged about why Spicer had made these false claims, adviser Kellyanne Conway said that Spicer had provided 'alternative facts'. President Trump continued to insist that he had the largest inauguration crowd in history and that news reports that questioned this were attempts to demean him and his supporters. He wasn't lying so much as confounding the meaning of reality. Veritocracy does not say that technocrats or philosophers should rule the world; it says that powerful people should not be confounding meaning but telling the truth and encouraging the truth to be told.

Veritocracy is a style of democracy guided and constrained by the aspiration to influence discourse and decision-making with truth as conventionally understood. Key to veritocracy is that politicians will allow themselves to be constrained in this way – not that truth leads directly to policies – the latter *would be* technocracy or epistocracy.

Another real-life example of what we mean occurred at the turn of the twenty-first century when Thabo Mbeki, the South African President, decided not to distribute anti-retroviral drugs to pregnant women in his country. Mbeki proclaimed to his Second Chamber that this decision would be seen to be justified if they inspected the controversy over the safety and efficacy of the drugs that was raging on the internet. But (perhaps unknowingly), Mbeki was not revealing what was true about the science: he was not reporting the view of that group most likely to tell the truth about the way the world worked, namely those scientists currently working on preventing the spread of AIDS. The view he was reporting could be found on the internet, but on the internet only: it was long past its 'sell by date' in the world of established science and could not be found in the respectable journals or respectable debating forums of science. But Mbeki, as a legitimate political leader, backed up by his party, was still perfectly entitled to make the political decision not to distribute the drugs. He could have claimed it was because they were too expensive with Western drug companies engaged in price gouging, or he could have said that he did not want to promulgate the idea that South Africa was an AIDS-ridden country. Both of these reasons might have justified a political decision not to distribute the drugs even whilst admitting that the scientific evidence in their favour was unquestioned. But in a veritocracy he should not have said the science was in doubt when it wasn't. Feeling bound to set one's decisions in the context of the truth in so far as it is known isn't technocracy or epistocracy because the decisions are not determined by the technical consensus. Mbeki could still have made the political decision he made in a *veritocratic* way, but in that case the justification would have been different: it would not have been based on an 'alternative fact' about the state of the science but political judgements – a legitimate political judgement – in the face of the science.

Slogans and truth

To repeat, we want to change the language of politics by introducing a new network of discourse that turns on integrity. Taking science as an object lesson, we see an immediate consequence for language itself: scientific language must be clear and unambiguous if it is to be open to criticism and that is a lesson for all decision-making language. Let us give one more example of what might change if the language changed in the way we want it to change – the use of slogans.

The extreme version of the opposite of the proper use of language in science is what we will call the 'polysemic slogan', which attracts supporters because it enables them to give whatever interpretation they want. The opposite – the idea drawing on the duty of clarity and unambiguity that provides the conditions for productive criticism in science – is the 'monosemic' slogan. The polysemic slogan works best when it is unanalysed – a rallying cry for all kinds of groups who might well have political ambitions that differ or even oppose each other and is directed to an audience beyond those with the expertise to criticize it (Figure 1.2); the monosemic slogan, on the other hand, has a logic that invites ready 'cashing out' into policies and provides for ready criticism, even though it will appeal to a narrower range of groups.

This matters most evidently at elections when citizens get to choose between competing policy platforms; they can sensibly try to do this only when the consequences of the different policy options are clear; empty phrases with broad but ambiguous appeal are the very opposite of this and, in this case, citizens cannot know what they are choosing. Monosemic slogans can be a rallying call or a ready reminder. 'Go to work on an egg' was the British Egg Marketing Board's attempt in the 1950s and 1960s to persuade people to eat eggs for breakfast; it was clear and harmless. 'Careless talk costs lives', 'Loose lips sink ships' and 'Walls have ears' were wartime reminders not to talk about operational details in public places, again clear and positive to reasonably fluent English speakers. 'Homes fit for heroes' is an example of a political slogan from the UK, harking back to the end of the First World War. Or there is our own clear and harmless slogan: 'the speed of politics is faster than the speed of science' (at least, we think it's clear and harmless).

On the other hand, what is meant by the slogan blazoned on Trump supporters' hats: 'Make America Great Again'? It seems to have only one meaning until one tries to imagine its application – until one tries to 'discharge it', or 'cash it out' – when it turns out to be quintessentially polysemic. It might mean do away with political correctness, it might mean abandon America's obligations to other countries, it might mean close America's borders, it might mean trying to stop black voters having any influence, and so on. It can mean 'vote for me and I will do whatever *you* think will make America great again', but with no information about what different groups of people want, nor how their dreams might be incompatible, nor how those dreams will be delivered. Polysemic slogans are a means towards the erosion of truth by embodying many alternative truths at once, all in the hands of the interpreters – a postmodernist's dream and, quintessentially, the opposite to the way science works, or should work.

The Brexit debate in the UK provides another example, with the campaign slogan of 'Take back control' proving more successful than the alternative of 'Stronger, Safer and Better Off'. But what does 'Take back control' really mean? Control of what and from whom? For example, on trade policy, did the UK regain control or surrender control over trade and travel to its nearest neighbours and biggest trading partners? Similarly, in the 2019 General Election, what was meant by 'Get Brexit done'? There was a clear sense that it meant 'sign a deal' but the content of that deal was never clear, and it turns out that the actual meaning was 'bulldoze through the legislation to leave the European trading zone before the intractable problem of the Irish border had been solved, causing financial losses, political chaos, and a potential return to civil war in Ireland'. Boris Johnson was a master of the polysemic slogan but as *Guardian* newspaper journalist Jonathan Freedland put it:

> For so long that was an argument made through the medium of abstract nouns: 'freedom', 'sovereignty', 'control'. But now reality is intruding.
>
> (Freedland 2022, *Guardian*, 3 Dec, 2022)

Using science as an object lesson for a veritocracy, polysemic slogans would be anathema, and all potentially ambiguous slogans used by respectable political parties would be 'discharged' – their means of execution and their implications spelled out. Of course, parties would disagree about whether the slogans had been accurately discharged but it is the aspiration that counts where the success of the end cannot be guaranteed; in a veritocracy it would be better to try to discharge a slogan and preserve truth than shamelessly use a polysemic slogan knowing that it promises nothing and that it does not constrain your subsequent actions. Politicians with integrity would point to ambiguous slogans used by other parties; the public would come to recognize the polysemic and vote for those trying to discharge their slogans: 'Truth not slogans!', as our proposed, veritocracy supporting, slogan might go.

Durkheim on professions and finding truth

One of the problems that societies face when it comes to truth, and one of the reasons for the world's weariness, timidity or just plain cynicism of academics when confronted with the idea, is that we no longer know where or how to find truth. The academic movement known as postmodernism has melted the meaning of words and even the meaning of experiments. According to many academics, 'the author is dead' – the meaning of words lost in the clamour of interpretation and re-interpretation. The same applies to experimental findings. So, the first problem we must solve if we are to make the case for veritocracy is a philosophical one: how to rescue truth from postmodernism; how to climb out of the 'postmodernist pit' as one might put it.

As hinted above with the story of the sage and the turtles, when differentiating veritocracy from technocracy we claim veritocracy has to be based on morality, with the *moral* culture of science playing a crucial role. As intimated in the Preface, the general idea might be thought of as going back to Emile Durkheim. Durkheim was concerned about the growth of 'anomie' in industrializing society as the traditional sources of stable social order were weakened by the social and geographical mobility that came with industrialization. He thought that one alternative source of moral order was to be found in the professions, with their self-imposed rules of conduct. He put this idea forward in lectures first presented in the 1890s, which were subsequently published in various editions as a book (e.g. 1958, 2019), with the title *Professional Ethics and Civic Morals*. In the Introduction to the 2019 edition, Bryan Turner explains that as France evolved in the late nineteenth century into an industrial society a gap emerged between civil society and the state that could only be closed by professional associations and a corresponding ethical system. So, Durkheim was introducing the idea that social values travelled 'down' from society into the formation of the professions but could also travel back 'up' from the professions into society. It is this latter insight that we are developing. In the academic domains of philosophy of science and science and technology studies, nearly all the emphasis nowadays is on the downward direction – how to bring social responsibility to bear on science – but we are arguing for the upward direction: we want the moral environment of science to inform and shape the moral values of the whole society.

Durkheim realized that professional ethics varied. He wrote:

> As professors, we have duties which are not those of merchants. Those of the industrialist are quite different from those of the soldier, those of the soldier from those of the priest, and so on. … there are as many forms of morals as there are different callings.
>
> (p5)

We are going to be concentrating on the generalized duties of scientists as-a-whole.[6]

What is the essential science?

In the Preface, and near the start of this chapter, we have invoked a science that is 'properly understood and executed'. We'll classify science, or descriptions of science, in three different places in this volume: here in this chapter, again in Chapter 2 and again in Chapter 4. The discussion of Table 1.1, is a first attempt to define science – attacking the long-standing problem of demarcation: what demarcates science from other approaches to knowledge? Philosophers seem to have given up on finding a solution to the problem, but we are going to provide one and in Chapter 6 we will use it to rule out certain activities that present themselves as coming under the heading of the science of economics. The key to the approach to demarcation taken here is that, though it draws heavily on the philosophical debate, it takes it that the solution has to be sociological, or 'institutional'. We can recognize sciences that correspond to the 'essence of science' by the formative aspirations of the institution – in the case of science, this is the search for correspondence truth. In so far as this goal is not being over-ruled by any other goal, then science is being done; where other goals take priority, then what occurs is no longer science.[7] In spite of philosophers leaving the field of play when it comes to the problem of demarcation, no longer believing they can find a set of necessary and sufficient defining criteria for science, if science is to be the foundation of veritocracy we have to do better.[8] We can do better by looking at the nature of institutions. We'll develop the idea later in this chapter and apply it to economics in Chapter 6.

Table 1.1 is a sketch of some descriptions of science available in today's world which will inform much of this chapter.[9] We'll go on to do some sub-classification in Chapter 2, covering the complexifying of the simple post-Second World War model and how descriptions '4–7' in Table 1.1 found space to grow out of descriptions '1' and '2'. We'll do some still more detailed classification as we go on, leading up to explaining the complexity of the interaction of science and the public. The main argument – that democratic societies should take science's aspiration to find truth as a foundation and object lesson – is a 'one liner' – almost a slogan – but it conceals a complexity in respect of the nature of science and its analysis that must be confronted if the slogan is to have any credibility, hence the multiple classification exercises.

Going back to Table 1.1, the first two entries are positive descriptions of science as it used to be thought about and is probably still thought about quite widely outside of sociology, history and philosophy. Descriptions 3–6 are negative in different ways, while description '7' moves us back to a positive position but one that takes into account much of what has been discovered about science since the 1960s and 1970s. We are arguing that the last description is the most fundamental *and that it is a valid description.*

Table 1.1 Some contemporary descriptions of science as an enterprise.

#	Description	Gloss
1	*The most perfect form of knowledge there is*	A set of procedures honed to produce the undisputable truth about the observable world as exemplified by its crowning achievements.
2	*A producer of wealth and understanding of the cosmos in peacetime, and victory in wartime, and a widespread ingredient of everyday modern life*	The central driver of economic development and the key to power. We hardly notice it in modern life but it gives us much of what we take for granted.
3	*A servant of professional targets, commercial and political pressures*	Nowadays a producer of publications, more than knowledge, subject to damaging internal competition, fraud, statistical malpractice, replication crises, misrepresentation of results for the sake of financial success, and subservient to governments' ambitions.
4	*An imperfect form of knowledge affected by subtle and not-so-subtle cultural influences*	Experimental and observational results can be differently interpreted so they can be affected just as much as other forms of artistic and humanistic production. This means science cannot claim elite status for its knowledge or expertise. Its outputs and procedures wrongly isolated from democratic accountability and self-conscious sensitivity to societal values.
5	*Just another literary construct*	An indefeasible 'postmodernist' philosophical description arising out of literary criticism and semiotics. Any activity of which the output is text can be analysed in the same way since 'the author is dead' and what counts is the interpretation of the text.
6	*A feature of culture which, like any other, can be interpreted to bolster political power*	It can be used as a pawn in the pursuit of political power like any other claim to represent the state of reality.
7	*'Craftwork with integrity': An institution uniquely defined by its collective aspiration, not always fulfilled in practice, to discover the truth about the observable world.*	Its practitioners are bound to aspire to tell the truth about their work (moral truth) as the necessary condition for any chance of fulfilling the collective enterprise (pursuit of correspondence truth) so science and scientists are bound to cleave to truth while pursuing their expert craft.

Description 7 is also the most important thing about science in the here and now. Science is more important than its findings and products because it is the upholder of the aspiration to tell the truth and at this point in history, Western societies need science for this reason more than for any other reason. And Western societies ought to be doing everything they can to stop commercial and political pressures as described under heading '3' from sabotaging this essential political and conceptual role: Western societies need science for its integrity if they want to preserve democracy. But because of the way economic and political life has been framed in the public's consciousness (heading '6'), we are in danger of destroying science and ourselves. Democracy contains the seeds of its own destruction, and it seems that, once more, we are rushing headlong to embrace it.

The most dangerous thing is the penultimate description; it is fatal when postmodernism leaks into the political domain. Its dangerous force rests upon science's reputation as described under '1' and '2' – a sure finder of truth and a provider of wealth. It is dangerous because most philosophers, historians, literary theorists or sociologists who study science in universities, pursuing the flaws in the perfection and efficacy of science, first under heading '4' and then under heading '5', have opened the eyes of dictators to the way truth can be manipulated even as it presents itself as a warrant for action. Once we believed that those days were over: with the defeat of fascism and its servant, eugenics, in the Second World War, and the demise of left-wing totalitarianism and its servant, Lysenkoism, in the Cold War, we thought we had finished with those days. But after a short golden age, which provided room for the licence to describe science under headings '4' and '5', we see the actual deployment of science as under heading 6. This is self-conscious in Russia's application of news as public confusion. At the same time, in the West, a stripped-down version of market economics is promulgated as something like a popular religion, giving rise to more and more inequality of wealth, and more and more of the political instability that history shows nearly always follows, but which is seemingly about to take the Western powers by surprise all over again.[10]

How can one support the claim that the 7th description of science is both valid and more fundamental than the others? After all, one can find science being practised in all the ways listed in Table 1.1 in today's societies, so how could one establish that one of them is the most fundamental? It seems like a very old-fashioned ambition, which is why it has been largely abandoned by philosophers, but it is a matter of thinking about the concept of science. This is not a proof; it is a matter of examining the form of life of science as we know it – a kind of philosophical and sociological 'revelation'.

Every description of science proper must include description '7' – you cannot take the pursuit of correspondence truth away from science and still have science. Thus description '1' is a declaration of the successful accomplishment of the aspiration described in '7'. And if '2' is right, it must be because at least a proportion of what science claims to have discovered corresponds to the way the world is, and this means the scientists must have been trying to find how it is. When we come to '4', '5' and '6', they allow that the pursuit of correspondence truth or the moral truth among individual scientists to be absent or at least subservient to other goals. The institutions represented by extreme versions of '4', '5' and '6' may pay lip service to the aspiration of science but are not really science *because* they do not rest on '7'. Instead, they are merely activities that present themselves under the label of science – elaborate confidence tricks in the most extreme cases and compromised or corrupted activities in the less extreme cases.

Any of the anti-truth or nil-truth descriptions may be correct descriptions of elements of the *profession* called science as it is found in today's world, and certainly may be the correct description of the *profession* of science as it may be found in tomorrow's world, but we know what we currently want from science: if we did not want a science that was aspiring to discover the truth then we should be honest about it and forthrightly embrace science fiction or power politics in its place.

But is it already too late for our political purpose to succeed? Is there too little left of 'the essence of science' in today's world for us to grasp it and pull ourselves up the tower of turtles to the political domain we want to establish? Is there anything left of the old institution based on truth-telling? To answer that question 'scientifically', we would need to start with a complicated and carefully executed survey of the activities describing themselves as science. But we can answer it a bit less scientifically: we know there is something left of description '7' both from fieldwork among scientists and from comparison with other institutions that appear palpably weak in terms of integrity when compared to science even when we include all its imperfections.[11] Our claim is that both conceptually and empirically, there is science in the world that fits the description under '7', that it is the essence of science as an institution, and that there is enough of it left in the world to act as an object lesson and the focus of a revamped civic education.

More features of an essential science

The importance of moral truth

We are saying that the crucial thing about science is its collective aspiration to find the truth about the observable world – what we call 'correspondence truth'. We have also mentioned what we have called 'moral truth' – this is the internal state of the individuals engaged in the collective enterprise because they realize that if they do not aspire to tell the truth about their own contribution to the collective enterprise, it will fail. The philosophical analysis of truth also deals with a 'coherence theory', but we are not concerned with coherence except in so far as consistency is a taken-for-granted aim of all theoretical discussions including this one. The two ideas of truth – moral truth and correspondence truth – are what we would encounter were we to study the activities of scientists in a field like gravitational wave physics.

The difference between moral and correspondence truth can be illustrated with examples of scientific change: when, prior to Einstein's and others' work, physicists claimed that energy and mass were conserved, they weren't lying; they were aspiring to tell the truth about how the world is. It has since turned out that they were wrong about the correspondence truth but their moral truth – their truth-telling intentions – remained constant.[12] Because science is always provisional, a veritocracy cannot depend on scientists telling the truth in the sense of correspondence truth; it depends on them trying to tell the truth in the sense of moral truth in order to maintain the aspiration to achieve correspondence truth.

In any case, too much concern about individual sciences, or even individual disciplines that do not fulfil the ideal, misses the point. Because we are building on the culture of science as a whole, and the ways in which that culture might feed into the culture of society more generally, we do not need the institution of science to be uniformly immaculate. What matters is that the intention to tell the truth remains a constitutive part of what it means to act scientifically – that is, moral truth remains a formative aspiration of science – and that democratic societies preserve the idea of science as an institution committed to discovering correspondence truth. Where these intentions align, then we can think of the science that emerges as what we can call a 'fulcrum science'. A fulcrum science is one that can support the working of veritocracy and, indirectly, be an agent of the change needed to bring it about. In sum, the hoped-for future depends on the institution of science continuing to practise moral truth in the quest to reach the often unattainable, correspondence-truth.[13]

This account of science is summarized and explained in Table 1.2. The formative aspiration of the institution of science is to discover the truth about the observable world (left column). This precipitates the formative imperative (middle column) that individuals engaged in the scientific work must tell the truth – this is a condition for the aspiration of discovering correspondence truth. Moral truth is far more robust than correspondence truth because it is an internal state so we can choose it, whereas we can only aspire to find correspondence truth. But even though moral truth is an internal state, it too is not always achievable because human frailty confounds intention. Mostly, this is inadvertent, the result of expectancy effects, confirmation bias and so on. The rightmost column of Table 1.2 lists methods that scientists use to try to link moral truth to correspondence truth. Once thought of as elements of 'the logic of scientific discovery', they are now better thought of as values internal to the institution of science designed to secure moral truth – designed to secure the path from choosing moral truth to accomplishing it.[14]

Table 1.2 How a fulcrum science works.

Formative Aspiration →	**Formative Imperative** →	**Procedures/methods to support formative imperative** ***some were once thought of as 'the logic of scientific discovery'***
Collectively aspire in expert groups to find **universal correspondence truth** in respect of the observable world	Individuals pursue **moral truth** by reporting findings honestly, accurately and clearly	Assiduous observation and experiment not preconception or revelation. Disinterestedness: e.g. use blind/double blind tests Corroboration/replication; falsifiability. Expertise and experience vital: controlled boundaries around small groups of experts *Locus of Legitimate Interpretation* (LLI) close to producers: fair criticism and response restricted to expert peers Arguments are aimed at convincing experts not citizens

Face-to-face groups

So important are both kinds of truth to scientists that they try to guard the boundaries of their professional interactions so as to allow in only those with the appropriate moral and technical socialization. They do their work in 'core-sets', which are small social groups.[15] Two examples drawn from the field of gravitational wave physics can be used to illustrate the working of trust in contemporary physics and will prove useful later in the book. The first is an example of personal trust.

For many years Russian scientists had claimed to have measured very good quality factors in sapphire crystals, but no one outside Russia could repeat the measurements. The quality factor, or 'Q', of a crystal is the length of time it will ring (like a bell) if set into vibration and a high Q would be important for the mirrors in the next generation of gravitational wave detectors. In 1999 a scientist, 'Checkov', travelled to Glasgow to show scientists how to make the measurement but even with his help they were unable to repeat it after a couple of weeks of work. Every indication was that the claim had always been false. But Checkov had left such a good impression of sincerity and conscientiousness that the Glasgow scientists, instead of declaring that they had uncovered a mistake or a fraud, continued to try. Eventually they achieved the desired result. In that case a subtle display of personal integrity led to the international acceptance of a scientific claim.[16]

The second example refers to a more general level of analysis. As Duhem pointed out, every scientific claim rests on a network of assumptions which remain stable only as long as scientists agree not to question them. In 2016, shortly after the announcement of the first detection of a gravitational wave, Barry Barish, one-time director of the project and soon to win a Nobel Prize for the discovery, was invited to present the details at a meeting of CERN scientists. A member of the audience remarked, correctly, that the whole story of the discovery depended on the assumption that gravitational waves travelled at the speed of light – something that had never been directly demonstrated. Another member of audience riposted that if questioning assumptions of this generality was allowed, they would soon 'undiscover' the Higgs Boson.[17]

That scientists must trust each other and that they must share the same assumptions or, at least, work towards sharing the same assumptions, means, as mentioned, that they mostly work in small, face-to-face groups (lines 4, 5 and 6 in the rightmost column of Table 1.2). This means that science can test and refine its claims by inviting continual criticism from trusted and knowledgeable colleagues. This does not guarantee to produce the pure gold of truth, but it does demonstrate, once more, science's aspiration to work towards the truth: it shows that scientists are willing to expose themselves to criticism and falsification. It also follows that, irrespective of the achievability of the aim of reaching the truth, scientists must aspire to write and speak about their claims in such a way as to minimize the scope for alternative interpretations. Scientists, in short, must try to be clear and unambiguous, continually fighting ambiguity however impossible it is to avoid the ultimate flexibility of interpretations – that too is one of the formative values of science. This is quite different to certain genres of art or literature, where a multiplicity of possible interpretations is considered a positive thing and vagueness and obscurity dressed up as performance can encourage a large following of interpreters.

Inside and outside science

Sometimes scientists must seek out competent critics in an active way, as when a scientific project, such as the search for gravitational waves or certain high-energy physics projects, has absorbed all the experts in the respective fields into the core team. In such a case they might actively set up a 'red team' from within their own project to work against the internal consensus as there is no one from outside the project who has the skills to do a good enough critical job.

This difference between the inside and the outside of research science is crucial if we are to understand the way science can work as an object lesson for other kinds of decision-making. It is also important if we are to understand how polysemic slogans and conspiracy theories based on bizarre interpretations of scientific claims can be taken as truth among the public, and why we must be careful about asking too much of public assessment of scientific results.

Within the small, closed groups in which pioneering scientific research is done, useful criticism comes from fellow experts.[18] Furthermore, useful criticism must begin with a fair representation of the position held by the criticized or it has no chance of changing the minds of other experts, only non-experts (line 6 in Table 1.2). Figure 1.2 is an *aide memoire* for this point about the difference between inside and outside: the inner disk represents the small group of experts with undistorted expert argument circulating within it, the intention being to approach ever closer to observable truth. But where criticism, seemingly of an internal opponent, is designed to be consumed in the outside world of non-experts, fair representation of the opposition's view no longer acts as a constraint.

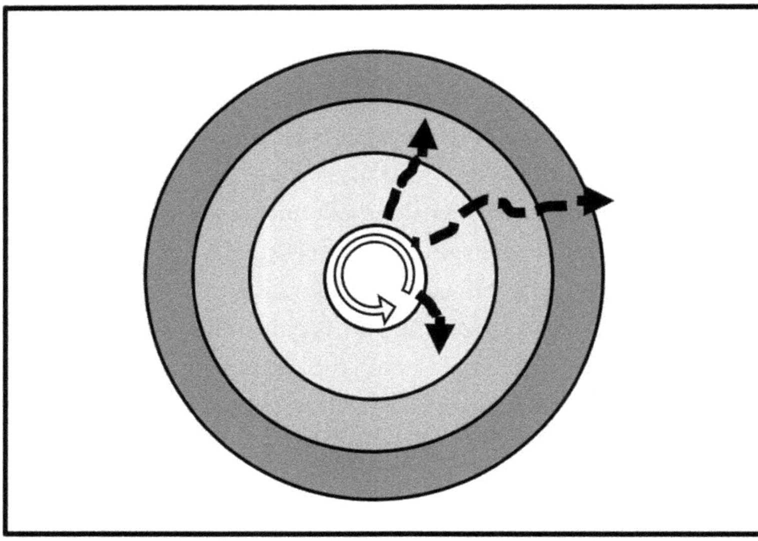

Figure 1.2 The target diagram (created by Harry Collins).

In the figure, black arrows are meant to be criticisms for consumption by the outside world – perhaps non-specialist scientists represented by the innermost annulus, or science administrators and journalists as we move out from annulus to annulus, or the general public beyond the outer annulus. These broken and distorted arrows represent the fact that the most effective criticisms designed for consumption by non-specialists are likely to begin with a simplified or distorted version of the specialist's view. The difference between this kind of criticism and science communication more generally is that the emphasis on criticism means that the simplified version that is presented is often a weakened or confounded representation of the specialist opponent's view. It turns out that, ironically, a version of the complex world of scientific expertise and argument that is found in the core of the diagram, with all its uncertainty, is experienced as far more certain as the distance from the research front grows and the complexities become invisible to the distanced view. The 'slogan' is 'distance lends enchantment'.

The difference between an interlocutor who is genuinely engaging with you in the inner circle and in the spirit of science, and one who is playing to a wider audience, is quite obvious to insiders, with the 'Science Warriors' of the 1990s providing many of the classic examples of this genre along with certain of the 'democratizers' (see next chapter). And, of course, criticism directed at a wide audience is the stuff of electioneering and political campaigning, which may include misrepresentation or deliberate confounding of both the opponent and the problem being faced. This kind of thing is integral to the creation of what we will call a 'ubiquitous science' (see Chapters 4 and 6).

Closed groups also safeguard the heartlands of science against the predations of social media, whose diffuse and often disguised users are not bound by the same moral aspirations and imperatives. It is no coincidence that social media is a fertile ground for conspiracy theories and fringe sciences. Social media creates an illusion of trustworthy intimacy not grounded in the face-to-face relationships that characterize core-sets. Information transferred via social media carries no obligation of moral truth.

Uniformity and universality

For completeness we will look at another feature of an essential science before moving on: it will figure strongly in Chapter 5 which deals with diversity, and we will add more analysis at that point. Uniformity implies that science's findings apply, in principle, to everyone across every culture; it follows from the formative intention collectively to discover correspondence truth. Universality, or universalism, implies that every contribution to science should be assessed without regard to the scientist's cultural (or racial) origins or gender since science is the same everywhere irrespective of origins. In principle uniformity and universalism are two sides of the same coin – you cannot have one without the other. Of course, we now know that science has traditionally ignored the world views and specialist knowledge of women, minorities and native peoples but we are going to suggest, in Chapter 5, a way that, with enough effort, it can emerge from these well-established problems and tensions.

Our aim should be to save science so we can save ourselves. If we scorn or destroy every institution that sometimes fails to live up to its aspirations, we will be left without anyone who will even *try* to guard the guardians: all that will be left is power. At the same time, given the traditional Western wellsprings of modern science and the long history of colonialism we must be sure that uniformity is not just a cover for exploitation. We'll go back to what uniformity and universalism mean in Chapter 5 but bear in mind that cross-cultural uniformity is regularly embraced in other cultural activities. For example, international sport would be impossible if, say, football, cricket and athletics were not driven by the same ideas and did not follow the same rules everywhere, even though there might be cases of corruption as in every institution.

Summary

In this chapter we have described what we are aiming for when we talk of veritocracy. We have offered up some descriptions of science as it is found in today's world and argued that some of these activities are not science at all. We have argued that the final description, description 7, is the most fundamental and that it is a valid description and that those who want to save Western democracies from extinction by establishing veritocracy should nurture that model of science and resist the pretenders. Towards the end of the chapter, we elaborate on the nature of a science based on truth. We now go on to look at the evolution of the understanding of science that gave rise to some of the descriptions of science listed above.

2

Evolution of our understanding of science and democratization

Introduction

Here we introduce a schematic account of the changes to our understanding of science that helps to explain why the admiration for science expressed under headings '1' and '2' in Table 1.1 could be supplanted by the more critical descriptions that follow it, and how the final positive heading arose. In particular we describe, under 'Wave Two', how a movement to 'democratize' science arose. Much of this will be familiar to philosophers, sociologists and historians of science.

Wave One

Once upon a time the argument for the trustworthiness of science would have had a different flavour. What we call the 'First Wave of science studies' reached its apogee in the years following the Second World War and was underpinned by the philosophy of science backed up by historical studies and, to a lesser extent, sociological studies.[1] The First Wave reached towards what Karl Popper described in the title of one of his books: *The Logic of Scientific Discovery*.[2] The crucial feature of the First Wave was to explain how it could be that science was the only kind of knowledge-making that stood above society and its influence. Science escaped from the dizzying implications of the sociology of knowledge, which holds that everything we think we know emerges from the circumstances of our early socialization and subsequent upbringing, each of which varies palpably from society to society. Science thus stood above both society and societies, generating universal knowledge from within a shared culture that transcended and cut across national and local cultures. In those days, history of science confirmed the logic of scientific discovery with its redescriptions of classic scientific discoveries and breakthroughs – the *crown jewels* of scientific discovery – while sociology of science explained the norms of science that were said to enable its transcendence.

Among thinkers about science there have, however, always been those pointing to aspects of science's provisionality and uncertainty, with Duhem and Fleck being important and Popper's 'swamp' metaphor being a well-known instance.

> The piles are driven down from above into the swamp, but not down to any natural or 'given' base; and if we stop driving the piles deeper, it is not because we have reached firm ground. We simply stop when we are satisfied that the piles are firm enough to carry the structure, at least for the time being.
>
> (Popper 1959: 94)

Nevertheless, if one wants to try to capture the spirit of the age, Wave One of science studies, bolstered by the success of radar and the promise of atomic power, was what was driving most analysts of science and science media personalities in the years following the Second World War. It is also what drives much contemporary science communication, with high-profile television series extolling the 'wonders' of science and rarely, if ever, exploring its uncertainties and failures.

To students immersed in this tradition, playing Popper off against Lakatos, the new idea put forward by Thomas Kuhn, that the history of science was characterized by paradigm revolutions, thus exhibiting its own cultural ruptures and incommensurable schisms, and having its own sociology, and sociology of knowledge, was a break with the past. If one was already familiar with the sociological interpretation of Wittgenstein's idea of form of life, it was easy to grasp because the notion of paradigm was form of life applied to the world of science.[3] But it was novel and dangerous to some of those who hadn't been so exposed. Indeed, it gave rise to great resistance from some scientists and philosophers, who wanted to go back to Wave One's understanding of science, embarking on what eventually became known as the 'science wars'. For example, an early 'science warrior', Israel Scheffler, described Kuhn's version of science as 'mob psychology'.[4]

Wave Two

For those who accepted Kuhn's insights, the 1970s saw a new approach to the analysis of science. This new approach affected sociology of science, history of science and philosophy of science, along with anthropology and ethnography of science, giving rise to what we have called the Second Wave of 'science studies' – although as is often the case when revolutions are proclaimed, many earlier precursors can be found.[5] What marked the application of the Second Wave out as different was its emphasis on science in practice – either through direct fieldwork or detailed archival research – and its determination to eschew both the rational reconstruction of science's successes and theoretical analyses of what its logical structure must be. Science studies grew rapidly and, although the inspiration might well have been *The Structure of Scientific Revolutions*, it was Wittgenstein's later philosophy that provided the detailed philosophical perspective for many of the early adopters.

From the early 1970s onwards, this new approach developed and grew, giving rise to its own paradigm revolution within science studies, even though Wave One thinking remained powerful elsewhere in the academy and the wider society, as it still does. Field studies of science in action flourished and anthropologists and literary theorists entered the fray with their seemingly more radical semiotic approach.[6] The work of historians was now often put to new uses, showing that the myth-like accounts of the crown jewels of science, re-told over and over again in the science textbooks, obscured the disputes and uncertainties of the time. Historical analyses now often showed that those disputes looked very like what goes on in the contemporaneous studies of the 'social construction' of scientific knowledge.[7]

Exemplary Wave Two insights

Interpretative flexibility and its consequences

That scientists work in small groups with relatively closed boundaries and huge stores of shared knowledge relates to how science progresses in the face of uncertainty. For instance, the experimenter's regress, which can arise when science is new or controversial, shows that unless you already know what the outcome of an experiment is supposed to be you cannot know whether that experiment has been carried out successfully and therefore you cannot know the meaning of a failure to replicate: does it show the original claim is wrong or is it a failure of experimental skill? The difference is hard to resolve because the transfer of the tacit skills needed to carry out a successful experiment is invisible. The consequence is that neither falsification nor positive replication can be used as a straightforward test for disputed claims, and arguments about such things turn into a search for proxies for competence and integrity on the part of the scientists; such proxies include personal trust and institutional affiliation.[8] In summary, Wave Two showed that experimental science was subject to interpretative flexibility and this allowed in the influence of culture.

The interpretative flexibility of experiments also makes it easy to understand the continued existence of 'fringe science' among the body of qualified scientists, and the existence of conspiracy theories among those more distant from the expert community: both are maintained through willingness to question the background assumptions and relations of trust that support the orthodox claims. As we move further from the mainstream, those wanting to dispute the conventional science are willing to question ever more basic assumptions and/or resort to ever more *ad hominem* arguments: the fringe scientists from inside the community may question, for example, esoteric but unobservable features of relativity theory, or the properties of sapphire crystals that they have never been in a position to measure themselves, whereas the anti-vaxxers and others from outside the community may question the veracity of institutions including the government, media or drug companies or simply dispute even the most widely accepted scientific views (see Figure 2.1).

None of this is difficult to understand in terms of the 'logic' of the situation. All groups of science consumers, including scientists themselves who have

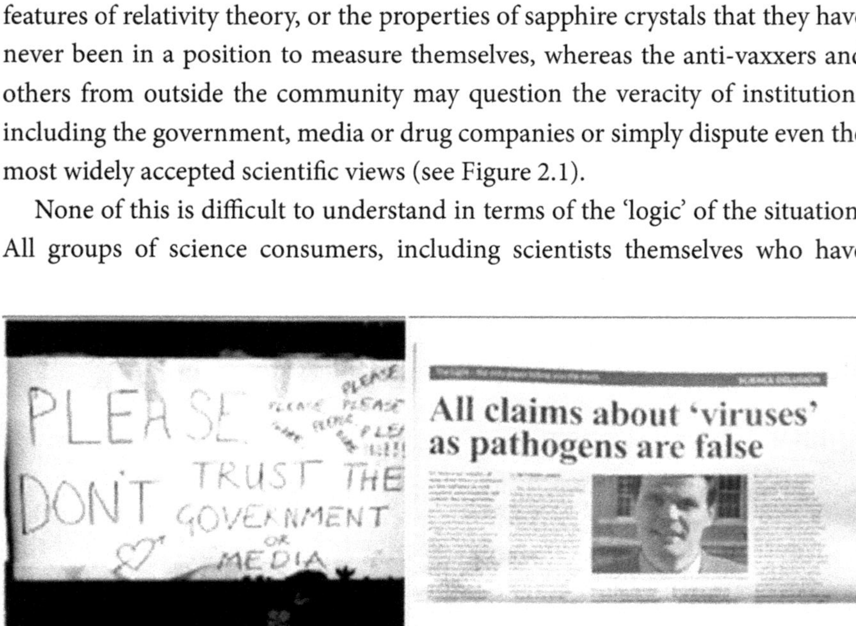

Figure 2.1 Anti-vax appeals in a shop window (photograph by Harry Collins).

continually to 'consume' science that they have not carried out directly, depend on hearsay for their knowledge of science. You, dear reader, have no direct evidence of the esoteric features of relativity theory nor of the veracity of the drug companies: you simply believe what you have been socialized into believing. Even if you are a medical scientist or medical practitioner, you almost certainly have no direct evidence of the effect of viruses on human health beyond hearsay; you just interpret what you see in terms of that hearsay. So, it is almost certainly the case that the difference between the views expressed in the shop window, as shown in Figure 2.1, and your views is a matter of which assumptions and which institutions you have chosen to trust.

But this is the important point: society works through shared trust in institutions, and this draws attention to the cure; the cure must be public trust in the institution of science, not individuals' trust in particular scientists and certainly not the supposed ability of individual citizens to make esoteric technical judgements. Unless we are members of the core set, or of a similarly esoteric group of experience-based experts, we simply do not have access to either the indicators of individual scientists' trustworthiness or the necessary substantive knowledge to make such technical judgements. Unless we are lucky enough to be the equivalent of members of the group who hosted the scientist from Russia who came over to demonstrate the right way to measure the Q of sapphire (and failed – as described in the last chapter), we must deal at the level of institutions. The fringe scientists and the conspiracy theorists differ not so much in their specialist expertises but in their failure to use their ubiquitous expertise in the same way as the rest of us; they just choose to trust and distrust different sets of institutions compared to the rest of us. This also opens up the way for the deliberate sowing of doubt by paying corrupt scientists, or encouraging politically motivated scientists, to promulgate sceptical papers about issues such as climate change or tobacco safety.[9] Though these contrarians may not be taken seriously within the core-group they can still have an impact on groups outside the core of the science (see, again, Figure 1.2). We will refer to such cases in more detail in Chapter 6.

Crown jewels, trust and policymaking

Wave Two also exposes another problem: the science that matters in most policymaking scenarios – the science that matters most to the public – is going to be provisional and contested, with different points of view and internal disagreements on public display. Unfortunately, the traditional rhetoric of science, and the way in which it is taught in schools, gives the impression that this is a pathological state of affairs, with 'proper' science reflected in the widely shared and iconic crown jewels of scientific discovery such as Newtonian physics' triumphant prediction of a new planet, Einstein's over turning of the Newtonian world view, quantum theory's huge but mysterious successes and, nowadays, false-colour pictures of the wonders to be found in the heavens. Though there may be the occasional near-miraculous event, or sighting, of this sort, most of the historical accounts of such events that are found in the textbooks and the popular media are well-meaning but false, taking no account of the uncertainties that always attend new scientific claims. As mentioned, Kuhn first pointed out that the textbook accounts of decisive experiments were wrong. This mismatch between the reality of policy science as science in the making and the image of scientific truth as quickly and decisively revealed is formed by retrospective accounting.

The Michelson-Morley experiment is a good example of retrospective accounting. Conducted in 1887, the Michelson-Morley (MM) experiment is said to have shown that the speed of light is a constant. It is said, in nearly all physics textbooks, to have reached a conclusive result by the time the measurements were completed. The results, it is said, gave rise to a troubling anomaly, which was not resolved until Einstein came up with his 1905, special relativity, paper explaining why the speed was constant.

This account of the MM experiment is a useful resource for *teaching physics* and appears in the popular literature. Thus, Stephen Hawking's *Brief History of Time* is a popular book found on the bookshelves of many households, even if, like the Latin Bible, no one understands it or even reads it. In it Hawking writes:

> In 1887 Albert Michelson and Edward Morley carried out a very careful experiment at the Case School for Applied Science in Cleveland. They compared the speed of light in the direction of the Earth's motion with that at right angles to the Earth's motion. To their surprise, they found they were exactly the same!
>
> (Hawking 1988: 20)

This is a typical crown jewels account of a passage of science and reflects nearly all the textbooks from which physicists learn their science and will be found in more readily accessible popular accounts of physics.

But more detailed historical reconstruction tells a different story. The MM experiment was meant to work as an Earth speedometer to indicate the speed of the Earth as it passed through the aether by measuring the changes in light speed. But no such changes could be measured so the experiment was abandoned as a failure. A series of improved experiments of the same type and with the same aim were then carried out in various places, the most elaborate being conducted in the 1930s. This one did find a small difference in light speed according to direction of travel and was even awarded a prize for doing so. All these experiments were subject to the experimenter's regress as Einstein's theory of relativity was not yet firmly enough established to settle the matter, though as it became more widely accepted so did the idea that a null result – i.e. no difference – was the correct one. It seems probable that no really satisfactory MM-type experiment which found constancy was conducted until the latter half of the twentieth century.

Though it might be useful for teaching physics, these crown jewels accounts are misleading as a representation of science for most purposes, including for policy advice. Science for policy is nearly always in the formative stage or is of a type that cannot produce exact answers. Science for policy invites questions about the special interests and competence of those who produce it and this can lead to demands that it should be interpreted, or even executed, by diverse groups who represent the different interests found among the public. If that is the case, then 'the problem of extension' remains to be solved.

Eventually a solid scientific consensus in respect of the theory of relativity, drawing on other kinds of experiments, resolved the experimenter's regress in respect of the Michelson-Morley experiment and set the scene for a backward reading of events. Henceforward, all MM-type experiments that found anything other than constancy for the speed of light are counted, in virtue of that fact, as incompetently performed while the very first iteration, abandoned by its instigators as a failure, is reconstructed as a decisive success. But the reading back all the way to 1887 reinforces the public image of the MM and of science in general as something with exact outcomes and sets up the conditions for disillusion and criticism of science as a producer of policy advice.[10]

Wave Two gives rise to democratization which aligns with populism

As the field of science studies developed, a strong interest emerged in showing that science no longer stood above society but was part of society; it was no longer uniform and universalistic but came complete with its own social influences. Initially, this work was often critical of overly technocratic forms of decision-making in which scientific knowledge was assumed to be superior to all other forms of knowledge and then used to exclude them. From the perspective of Wave Two, the demonstration that science was as socially constructed as any other form of knowledge meant that it should not be used in this authoritarian way. Instead, more inclusive and reflexive forms of decision-making were needed in which wider concerns were given a greater say. Thus, as Sheila Jasanoff has put it:

> Many STS scholars think that the institutions, practices, and products of science and technology should be characterized in new ways not only for the sake of descriptive adequacy and analytic clarity, but also in order to reorder power relationships: for example, to make the exercise of power more reflexive, more responsible, more inclusive, and more equal.[11]

Whilst this concern with the misapplication of scientific knowledge and its authority is entirely reasonable, and new ways of using science within decision-making might have been overdue, it also led to the much more dangerous claim that science not only did not, but *should not*, stand outside of democratic politics.[12] This movement was brought to a focus by Sheila Jasanoff and Bryan Wynne who developed wide followings. We'll refer to these two and the people they influenced as 'the democratizers' and the widespread tendency to which they gave rise this whole tendency, as 'democratization'.[13]

Here was a clear divergence in attitudes about what follows from having discovered the complex ways that social influence infiltrated scientific research. One could take 'Fork A', concluding that we must be more vigilant to try to secure what we value about science, or, like the democratizers, one could take 'Fork B', concluding that now we know that science is pretty ordinary we should treat it as just another feature of political life and grant it no special status or privilege.

The old debates about the relationship between science and democracy going back to Dewey and Lippmann were revisited and interpreted accordingly, with those taking Fork B seeing Dewey as the hero and Lippmann the undemocratic villain.[14] For those following Fork B, science was an opaque body of knowledge controlled by an elite with unjustifiable power, based not on election to office but chosen through the procedures of closed-door societies belonging to the esoteric world of fellow professionals. As such, it must exist in a state of tension with democracy as it threatens to undermine and exclude the concerns of ordinary citizens. This new way of looking at the world, in which science is always the problem and more participation is always the solution, was symbolized with the invention of the term 'lay-experts', intimating that ordinary peoples' views about scientific matters had as much right to be heard as the views of professional experts since the experts were no longer immune from the social influences and self-interest that affect the rest of us. Thus, Sheila Jasanoff, emerging among the most successful of those championing the democratization of science, wrote:

> The presumption in democratic societies is that all decisions should be as far as possible public; it is the exceptions that require justification. … expertise is constituted within institutions, and powerful institutions can perpetuate unjust and unfounded ways of looking at the world unless they are continually put before the gaze of laypersons who will declare when the emperor has no clothes. (2003: 397–8)

Jasanoff would go on nearly twenty years later (at a talk given 2022), just at the time when one might think that the unfolding of events under President Trump had made the danger of democratization more evident than ever, to justify the idea in the context of political philosophy:

> Kant represents the enlightenment in demanding that people make judgements for themselves and not take the word of authorities. Immaturity would be to trust what a book, or a spiritual director, or a doctor says; instead, you're supposed to be exercising, in some sense, your own judgement. Of individual enlightenment and maturity to exercise your reason to weigh the decisions of the world? One thing that's happened is that we've written, de facto, a new constitution for modernity which was not in the original wording of the revolutionary era constitutions of the late eighteenth century. Today … we have masses of areas … in which we delegate our authority to experts precisely to tell us how to behave. … we grant them epistemic authority to know for us how we should be exercising judgement, and then when we refuse [eg to accept the advice of people like Anthony Fauci, medical advisor to the US President] it becomes a problem because people are not trusting science because they should have been trusting science. So, this is a huge evolution in 200 or so years, away from a view of enlightenment that said it was a good thing to doubt, it's a good thing to trust your own judgement, to a position that we're in now, where distrust of institutionalised authority is seen as a problem in and of itself. So, I think that even to begin to get to grips with the state we're in, we have to begin by recognising that these expert institutions that we've created are not simply telling us the state of the world in an unmediated fashion so that we can trust

> those claims in a direct way, instead they are exercising authority that is political ***in the very same way*** that our elected bodies are political, or our politically appointed bodies are political.
>
> (Jasanoff 2022 [around 12.00 minutes in, our stress])[15]

The practical meaning of this kind of claim had been made clear in the democratizers' sparsely documented, but clear and explicit-at-the-time, support for the revolt against the deployment of the Measles, Mumps and Rubella (MMR) vaccine. The revolt had been triggered by the press conference to announce Andrew Wakefield's notorious, 1998, article in *The Lancet*, and during which Wakefield went beyond the published paper to claim that the results could be interpreted as indicating that the MMR vaccine could cause autism. Many parents, experiencing the agony of seeing their child exhibiting the symptoms of autism shortly after administration of MMR vaccine, and supported by newspapers with an eye for a 'balanced' story, were sure that the temporal relationship was a causal relationship in spite of epidemiological studies of populations before and after the introduction of the vaccine showing no relationship. As a result, vaccination rates fell below those needed to maintain herd immunity, putting vulnerable children who could not be vaccinated, plus all those whose remained un-vaccinated as a result of parental choice, at much greater risk of measles, a disease which is known to be both highly infectious and capable of causing great harm. Despite this, most within STS either said nothing or used their platform to defend the rights of those opposed to the vaccination. In January 2024 it was reported that measles infections had increased thirty-fold in Europe consequent on vaccine hesitancy growing out of the initial revolt,[16] but there has been no reflection on the matter by the democratizers.

The other principal democratizer, Brian Wynne, is particularly well-known for his study of Cumbrian sheep-farmers and his work criticizing the 'deficit model' in the public understanding of science. Although not the source of the term 'lay expert', Wynne's study of Cumbrian sheep farmers has been hugely influential in the argument that lay people may object to science for perfectly rational reasons and their scepticism should not be interpreted as the product of ignorance, and hence something that would disappear if only they understood the science properly.[17] In the case of the sheep farmers, Wynne shows that they had knowledge of sheep ecology that should have been used in scientists' analysis of post-Chernobyl sheep farming in the Cumbrian fells and that the rejection of this by the scientists, combined with frequent changes to the scientific advice that was issued, contributed to the farmers' sense of scepticism. Unfortunately, the paper does not recognize that most of the knowledge and experience the farmers relied on in making these judgements was not the everyday lay knowledge available to anyone, but a specialist expertise acquired through years of experience of farming sheep in a unique upland setting. In other words, although the argument is framed as being about the 'public uptake of science' and makes much of the contrast between the 'ordinariness' of the sheep farmers and the elite character of the government scientists, the key protagonists – the sheep farmers – are not lay persons at all but an equally elite body of experience-based experts (elite, not in the sense of rich and powerful but elite in respect of their technical understanding, and with well-defined and impenetrable boundaries).

Despite this, the 1992 paper is presented as, and continues to be widely read as, exemplifying the capacity of lay people to engage with science in complex and sophisticated ways. Similar ambiguity can be seen in the papers that do introduce the term 'lay expert' (Arksey 1998; Epstein 1995, 1996), both of which concern medical settings in which the 'lay experts' are highly motivated patient groups who have proactively created a sophisticated body of knowledge about their own condition and made extensive efforts to engage with the medical profession. Again, whilst these patient groups are undoubtedly not 'elites' in the way that doctors and scientists are, they are not 'ordinary' lay people either but elite groups in terms of their experience.

This lack of critical thinking about who or what 'lay experts' are feeds into the way the ideas are used later on, and how many practitioners of science studies then respond to public controversies. In short, anything that sounds like a deficit model explanation – that is, one that implies lay people might be making a mistake or lacking relevant expertise – is rejected instantly with the author often being criticized for their technocratic tendencies. Instead, the default position is that rejection of lay interventions is a failure of democratic politics rather than an appropriate exercise of democratic authority, accepting advice from experts in a way that is needed when making technical decisions in democratic societies. On this basis, Wynne, Jasanoff and others in STS endorsed the view of the parents in the MMR controversy, and, as can be seen in the extensive quotation from Jasanoff above, justify the view of the anti-vaccine position that is currently found well represented among the public. The recent COVID pandemic has produced another opportunity for the safety of mass vaccination programmes to be debated, and whilst there is a general sense that most people in science studies support the vaccination programmes, there has been little explicit discussion of vaccine safety or why and whether the COVID case differs from the MMR controversy. In other words, why shouldn't the public's view of COVID treatments be sufficient to justify a precautionary approach, just as it was at the time of MMR?

The revolt against MMR and the way Wynne, Jasanoff and others within science studies supported the parents who were rejecting vaccination are part of what triggered the 2002 paper by Collins and Evans defending the idea of expertise. It is in that paper that the history of science studies is first divided into two 'Waves' with a new Third Wave based on expertise and experience being put forward (see below). The distinction between the problem of legitimacy and the problem of extension was also introduced. The problem of legitimacy refers to the difficulty for the acceptability of scientific institutions among the public created by too technocratic an imposition of science and how this might be ameliorated by bringing the public into science. In contrast, the problem of extension is how to retain the notion of scientific expertise if the right of holding a scientific opinion is *extended* to the general population; if it is, then science merges with democracy and the idea of specialist scientific expertise is lost.[18]

The contemporary rise of populism has increased the importance of this debate still further. In the current context it is easy to see that if we want to maintain a system of checks and balances in Western democracies then at least some of them must have power born out of their technical expertises and this means they must be elites in terms of the recognition of that expertise. The authority this grants to science strikes a discordant note with conspiracy theorists and anti-vax groups, and these might well support the sentiments expressed by Jasanoff and others in science studies (perhaps to the latter's embarrassment). But the idea that there is an inevitable conflict between elite experts and democracy is easily seen to be false. For example, consider the way things work in respect of the law, which is also one of the important checks on the political elite, and for which no one is seriously arguing no specialist training or expertise is required. Of course, this is not to say that populists will not try to undermine or capture the court system. In the UK, when Boris Johnson wanted to prorogue Parliament to push through his Brexit legislation with minimal debate, the judges of the Supreme Court prevented him by declaring the proposal illegal. Indicatively, a Brexit-supporting, right-wing newspaper, *The Daily Mail*, dedicated a front page to a photograph of the principal judges with the caption 'enemies of the people' (4 November 2016). In the United States, of course, the rule of law seems in danger because of the influence of the people, via the elected President, on Supreme Court appointments. Both the *Daily Mail* and the evolution of the US Supreme Court align with the democratizers' sentiments that elite institutions should follow public opinion not their own expertise. In contrast, we would argue that both these examples show the importance of independent, elite institutions in protecting democracy from those concerned only with advancing their own interests.[19] The democratizers remain silent on the relationship between legal expertise and scientific expertise.

Wave Three

Wave Three was a reaction to the democratizers' over-interpretation of Wave Two in terms of the erosion of the political significance of science. Wave Three's problem is how to justify science as a source of legitimate expertise in the light of Wave Two's findings. Wave Three focuses on the culture of science and what makes it distinctive compared to other social groups and institutions. The least controversial Wave-Three move is to shift the focus away from truth and onto expertise in general and then scientific expertise in particular. Scientific truth is at best a long-term business, but expertise can be analysed and, to some extent, evaluated, in the short term. A number of new concepts arise out of the idea that expertise was a matter of socialization into an expert community.

Wave Three gives rise to a typology of expertises that includes both specialist and ubiquitous expertises.[20] Within the specialist expertises, the distinction between 'interactional expertise' and 'contributory expertise' has proved particularly useful: interactional expertise refers to fluency in the 'practice language' of an expert group, whereas contributory expertise involves practical skills too. Separating the language and practice components of a form of life has led to a number of methodological and substantive insights, all linked by the realization that the language element of a culture is far more significant than has been recognized in typical academic philosophy, sociology and psychology.[21] Thus, natural language speaking is a ubiquitous expertise and so is the moral code of a society – they both belong to all the members of a society who acquire their foundations without conscious effort.

The idea of specialist expertises as based on extended socialization creates a bridge between scientific expertises and the experience-based expertises of social groups engaged in relevant areas. On the other hand, by distinguishing specialist from ubiquitous expertise, Wave Three recognizes the difference between the expertise of elite groups, in the sense of small numbers and well-defined boundaries if not high status, and the expertise that is widely distributed among lay people.

More contentious elements of Wave Three are the re-conceptualization of what was once thought of as the logic of science as a set of values that make science work as an institution (as found in the rightmost box of Table 1.2) and the strong emphasis on the value of truth as fundamental to science and to society. It is Wave Three that gives rise to the seventh description of science in Table 1.1 of Chapter 1.

Wave Three argues that policymaking must turn on an historically more accurate model of scientific discovery than is found in Table 1.1's first and second description, especially those elements supported by the crown jewels. The analysis of expertise and the way it fits into the moral world is the way forward. The moral analysis of science is what makes it the pre-eminent provider of technical advice. But assuming they are honest about the advice they are being given, and the way they have assessed it, elected politicians, not scientists, should make the final policy decisions – the alternative would be technocracy. Indeed, politicians should make the decisions even when the science is pretty-certain, so long as they explain how they are making them in the light of the scientific advice. If it appears they are going against established scientific wisdom, then it is especially important that the explanation is clear.

There are technical disagreements in the early stages of any science; this is inevitable if the internal criticism is working properly. What the politicians should be doing is leaving scientists to do their best to settle their disagreements in the political timeframe but within their tightly guarded competence networks, listening to what they say, making an assessment (aided by social scientists it is claimed in Collins and Evans 2017b), about how strong the scientific consensus is, and then making clear just how the science has fed into their political choices. But for that to work, politicians and ordinary citizens must agree that the institution of science is sufficiently trustworthy to leave at least that much of the input to policy – the advice on matters of science – to the institution of science. It is not that the findings of science are *necessarily* more true than other forms of knowledge but its judgements about what to believe are reached in a morally superior way, so it is the way to bet. Or, in more commonsensical terms, if you want someone to tell you the truth about the world the best place to start is with a group committed to telling the truth.[22]

A philosophical parallel: Scientific value analysis?

Parallel to the changes discussed above, a movement has developed in philosophy, which we will refer to as 'scientific value analysis (SVA). It is interesting to compare it with both the democratizers and what we are proposing. A summary text is Kevin Elliot's (2022) *Values in Science*, while one of the principal instigators is Heather Douglas with indicative books in 2009, *Science Policy and the Value Free Ideal* and 2021, *The Rightful Place of Science: Science Values and Democracy*, which includes commentaries such as the one by Daniel Steel, referred to below.[23]

The analysis focuses on the way that societal values do and should influence scientific debate. In terms of Figure 3.2 (see next chapter), SVA does not consider the upward pointing arrows. SVA makes a useful distinction between values internal to science and non-epistemic values. There is no incompatibility between SVA's values internal to science and Wave Three's, even if the content may not be identical. Thus Table 1.2 sets out values internal to science related to truth with the rightmost box urging the recasting of the old 'logic of scientific discovery' as a set of values rather than rules or mechanical procedures. The same idea is put forward in Collins and Evans's (2017b) *Why Democracies Need Science.*[24]

An important element of SVA is drawing attention to the way 'inductive risk' should affect features of science that are often thought of as strictly 'internal'. For example, what level of statistical significance should warrant the publication of a scientific claim? The SVA answer is that it depends on the extent to which the use of advice based on that claim in the creation of policy would expose the public to risk. The MMR example, where Wakefield's claims were based on results with almost no statistical significance or scientific warrant, nicely illustrates the point; it would have been incomparably better if those results had been withheld until they had been far more solidly proved (or disproved). This accords with Wave Three's view about how policy should be made, with part of the job of 'The Owls' (see *Why Democracies Need Science*), being to advise policymakers about the strength of scientific consensus in respect of a claim. Furthermore, the compatibility is emphasized in that Collins puts forward a similar idea in his 2004 (*Gravity's Shadow*) description of the early attempts to detect gravitational waves. Collins offers a thought experiment (pp781–2) imagining that certain claims to have detected weak fluxes of gravitational waves impacting the Earth would probably have been assessed differently if there had also been a credible view that the negative effect of power lines on human health was enhanced by occasional gravitational waves.

But versions of SVA are more ambitious than this, suggesting that science ought also to be influenced by wider societal values. For instance, Douglas worries that scientists can be damagingly isolated from these values through their scientific socialization:

> Many of the most central values scientists hold are those they developed whilst training to be scientists, and this alone can create divergences between the scientific community and the general public… Thus, even a demographically diverse scientific community may still hold values that clash with the broader public's values.
>
> (Douglas 2009: 122)

The trouble is that if the philosophical claim is that the broader public's values should influence science there is nothing to stop the argument being used by the leaders of malign societies (or their advisors). As we will see in Chapter 5, Sandra Harding recognizes this, citing Nazi society as an example (Harding 1995a: 25). And, of course, this whole book is an attempt to stop the influence on science of the values emerging from what American society is threatening to become. Instead, it is arguing that science's values should be preventing America and other Western societies turning into the malign societies they are becoming. Again, this is an illustration of the tension between the left-hand panel and the right-hand panel of Figure 3.2 in the next chapter, with emphasis on the left-hand panel offering comfort to populist leaders trying to subvert science's constraining influence on their power to define the world as they prefer.

Though the initial discussions of the way values are invested in science accord with what is being argued here, the extension into the acclaim for the influence of societal values on science moves SVA in the direction of the democratizers. Subjectively, the democratizers seem far more sceptical of science than the proponents of SVA, drawing them to act almost as though they were involved, if not in full-blown war against science, at least as though acting out the tension between C. P. Snow's 'Two Cultures'. SVA, on the other hand, appears broadly in favour of science, trying to improve it, not oppose it. One cannot imagine SVA endorsing either Wakefield or the parents who supported him against the evidence of epidemiologists. Nevertheless, SVA needs to find a way to distinguish between the values of the public that it would endorse as an influence on science – such as non-exploitative treatment of experimental subjects – and those it would not endorse. One's sense is that this latter list would be a long one and include many of those that would be found in an America that elected a Trump-like President. But this is far from clear at the moment.

In the 2021 volume, Douglas does set out her view of how public values might be incorporated into science. She imagines discussions between disagreeing scientists and/or citizens, asking each other what evidence would change their minds, and ruling out of consideration those who cannot provide an answer. But such ideal conversations do not have any bearing on the real world of conspiracy theories or even the scientific fringe; they depend on the existence of the equivalent of the university seminar room populated with well-intentioned and well-educated debaters all sharing a belief in the scientific method described in Table 1.2.

Steel in his, 2021, commentary points out that Douglas does not specify the way participants will be selected for her favoured deliberative forums. Indeed, it sometimes seems hard to find these well-ordered deliberative forums even within the world of academia; after all we cannot get the democratizers to admit they made a mistake over their analysis of measles mumps and rubella vaccine – seemingly no evidence or analysis will lead to them admitting this mistake in public. One might say of Douglas's well-worked out philosophical analysis that it lacks the scars of contact with more than an idealized range of arguers. Here, as we explain in the Preface, we are dealing with the survival of respect for science in general and worry that these imagined seminar debates will give a licence to citizens of all kinds to think they can contribute meaningfully to the discussion of the substance and legitimacy of science whereas real experts are limited to small elite groups.

Table 2.1 is an attempt to compare the relationship between science and the public as preferred by the democratizers, scientific value analysis and the Third Wave.

Table 2.1 Comparison of three approaches to science and the public.

Dimension of Comparison	democratizers	Scientific Value Analysis	Third Wave
View of science	Antagonistic	Positive	Positive
Contribution of public to science	Same as to politics	Important to align science with public values	Analyse on a case-by-case basis, experience-based expertise being necessary at the research front
Contribution of science to public values and politics	None	Not considered	Vital if democracy is to be saved
Melding of public and scientific expertises	Public's democratic choices dominant	Scientists should absorb public values	Virtual diversity hands responsibility to scientists to understand and work with experience-based experts and choose which societal values to bring in

Summary

In this chapter we have provided a historical sketch of how the models of science at the top of the list set out in the first chapter (Table 1.1) evolved into the other descriptions. We discuss our attempt to establish a Wave Three of science studies, based on an analysis of expertise and of the moral basis of the institution of science linked to seventh description of science in the list. We described the movement, which we call democratization that grew out of Wave Two and then we described the more subtle parallel in philosophy, which we call scientific value analysis. Table 2.1 summarizes the three approaches. We now go on to describe the workings of societies in terms of Wave Three's analysis of expertise.

3

The fractal model of society

Knowing what we mean by science we can now locate sciences on 'the fractal model of society'. Going back to Durkheim's insight about the possible role of the professions, we modify and generalize it into domains of expertise which are less formal than professions and rarely come with a certificate of completion of apprenticeship. These informal domains still have concrete conditions of entry, namely socialization into the culture, but socialization is only rarely accompanied by formal qualifications. These social domains can be pictured by using the idea of a fractal as a metaphor – see Figure 3.1. This figure will be used

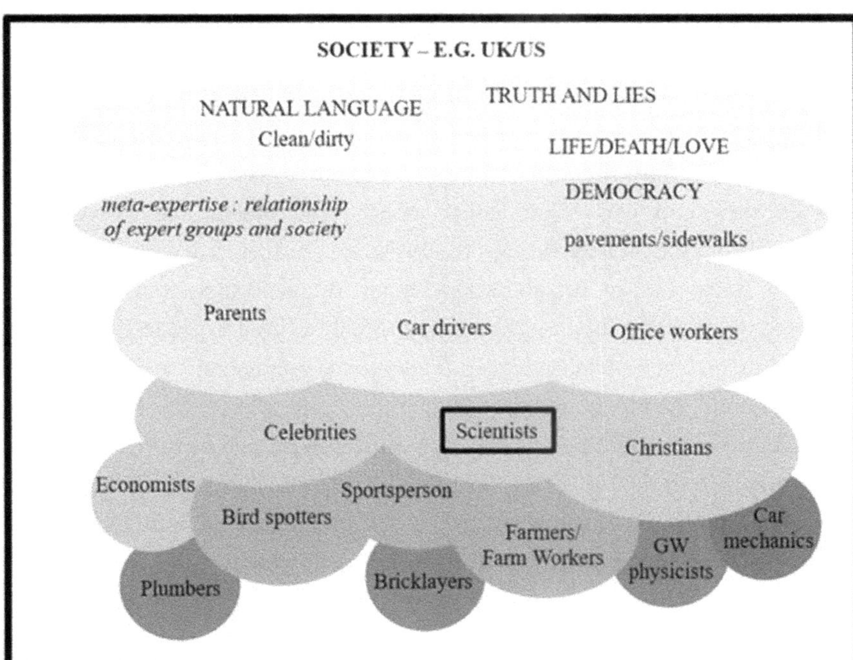

Figure 3.1 The fractal model of society (created by Harry Collins).

as a motif throughout the rest of the book, modified as appropriate to represent the various strands of the argument.

An expert community under the fractal model of society can be of any scale from nation to small groups of specialists or hobbyists; they overlap and are embedded within one another as illustrated in Figure 3.1. At the top of the fractal model are the groups coextensive with the state (or tribe) in which all citizens share the 'ubiquitous expertise' that signifies membership of that society. The very top group represents the basic constituents of social life including a sense of consistency and truth develop in babyhood as the rudiments of language are acquired, with carers having to be constant in their associations of sound and object if words are to be learned. These basics are almost certainly common to all human societies giving rise to what can be thought of as a still higher-level humanity-wide fractal. Below this is an oval of similar size representing the specific society-wide, 'ubiquitous expertises', characterizing individual societies. They include the way the basic constituents are instantiated in different societies, with different natural languages, different moral codes and different distinctions between clean and dirty. Important for our analysis is the meta-expertise found at this level which guides citizens in how to choose between the social groups of experts found lower down in the fractal: when they are sick, do they choose to go to a doctor or to someone who can divine the witch who has applied the curse and so on? There are many other less basic differences between societies such as the space maintained between persons as they pass each other on the pavement or find themselves pushed together in crowds, the relationship between men and women, different supernatural beliefs, the assumptions about proper political behaviour – the thing we are trying to influence – and so on. It is citizens' or members' possession of these ubiquitous expertises that make one society different to another. Most of these expertises start to be acquired in 'primary socialization', in the family though some, such as political understanding, start to be acquired later in life. Cascading below this top level there are ever more specialist expertises of which science is the example on which we are concentrating. The expertises pertaining to smaller groups are acquired during the later stages of socialization that take place after the individual enters institutions such as schools, universities and workplaces. This is 'secondary socialization' though we will sometimes use the term 'tertiary socialization' for what happens in universities and workplaces.

All expertises are acquired by socialization within a group – there is no distinction between culture and expertise. The expertise needed to be a plumber is acquired through secondary/tertiary socialization among plumbers; the expertise needed to be a gravitational wave physicist is acquired through socialization among gravitational wave physicists, though this will be embedded in the higher-level socialization into being a scientist. And to be an expert astrologer, one must spend time in the community of expert astrologers. All acquisitions of expertises and of cultures, whether large or small, ubiquitous or elite, are similar, and this justifies the use of the fractal metaphor. The implications of the fractal metaphor will be further expanded in Chapter 5, where it will be used as a basis for our discussion of what we will call 'virtual diversity'.

What concerns us in this chapter is the interaction between the ubiquitous expertise that all members of a society possess and the expertises acquired in secondary (and tertiary), socialization. Primary socialization will equip the citizen with, among other things, native language and basic moral values. Subsequent specialist socialization cannot take place without the language and the foundations of moral understanding such as truth. The left panel of Figure 3.2 has a forest of downward pointing arrows emerging from the top level and influencing every sub group with the many features which define a society, with language and a basic understanding of truth being fundamental. In the long run, any changes in these societal values will lead to changes in all the smaller groups that inherit them. In some cases, these might be changes for the better as, for example, when beliefs about gender equality change, leading to changes in the way women are treated in a wide range of different settings. In other cases these changes are damaging, such as the changes in the professions that have come about through the spread of neo liberalism in Western societies, and this kind of thing is what we see happening with the erosion of truth under Trumpism. If truth loses its meaning in society in general, this will be reflected in the conduct of smaller groups; ultimately, language will everywhere be replaced by force.

Here we are imagining what Durkheim might have thought about today's society by emphasizing the right-hand panel of Figure 3.2, where the arrows go back up from the sub groups to the top level. The idea is that, as secondary socialization gives rise to subsidiary institutions such as the professions, these have the potential to feed influences back to the ubiquitous expertises, affecting the entire society. We could say that the entire book turns on the relationship between the left-hand and the right-hand panels of Figure 3.2. Both downward and upward influences are essential features of society, but the left-hand panel is being more and more analysed and explored in today's sociological and

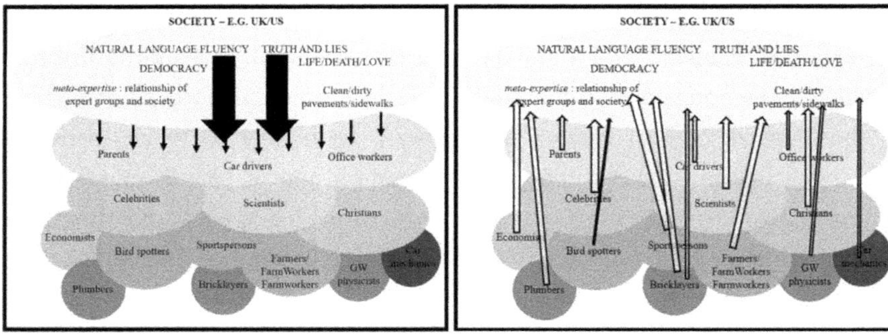

Figure 3.2 The fractal model showing downward influence of existing ubiquitous culture on subgroups (left panel) plus the upward influence of sub groups on society as a whole (right panel) (created by Harry Collins).

philosophical circles, while the right-hand panel is more and more ignored in modern treatments of knowledge.

Here we are going to stress the role, actual and potential, of the institution of science in stabilizing society's respect for truth. Science's cross-societal uniformity allows it to influence something as basic to all societies as the role of truth. At the same time, we are criticizing the over-emphasis on the downward influence of the values of particular societies in the analysis of science in general, and in the practice of certain sciences such as economics, much of which, we will argue in Chapter 6, fails to exemplify the formative values of the institution of science because of its pathological integration with public and political discourse. This has reached the point where in Western societies such as the United States and the UK it has come to exemplify the negative side of the sixth description of science listed in Table 1.1: *a feature of culture which, like any other, can be interpreted to bolster political power.*

As Durkheim has fallen out of fashion in the social sciences, with analysts losing interest in 'treating social facts as things' and turning their attention to the places where groups overlap or are ill-defined – a change of attention from the central formative characteristics of social groups to the fuzziness at the edges – the thrust of social and philosophical analysis of science has been with the left-hand panel. The traditional descriptions of science have it standing outside of society and uninfluenced by social forces but, for most of the last half-century, the social studies of science have focused almost exclusively on criticizing this traditional view, either stressing or advocating the downward pointing arrows: the influence of, or what ought to be the influence of society upon science; some of this was discussed in the previous chapter.

In pursuit of veritocracy it is going to be argued that, in contrast, science does, or should, provide a particularly important influence on the wider society. These arguments will be developed as the chapters unfold. To anticipate, some of the reasons for the special importance of the upward influence of science are:

i. because liberal democracy, insofar as it depends on the idea that my freedom stops at the end of your nose, requires the detailed *findings* of science to explain where the end of your nose is located (think, for example, about vaccines)[1]
ii. because scientific decision-making when it comes to establishing those detailed findings is the best *role-model* for political decision-making under uncertainty
iii. because science is one of the vital checks and balances needed in pluralist democracies
iv. and, most important of all, because science is the most important institution bearing on the most fundamental value of society, a taken-for-granted respect for truth, without which there cannot be values, facts or a human existence – there can only be dystopia

Going back to the fractal model, on the matter of the definition of a social group, anything counts as an expertise so long as the members of the associated group could, in principle, pass or fail in an Imitation Game, indicating that there is something called a culture associated with the group. This differs from Durkheim's idea where professions were likely defined by apprenticeships and, often prestigious, formal qualifications. Some of the sub groups in the fractal model will be like this but the Imitation Game leads to a far more general idea of 'group'. An Imitation Game is like a Turing Test but with a person doing the pretending rather than a computer.[2] An expert judge cross-examines a hidden genuine expert and a hidden person pretending to be an expert and tries to work out who is who. Thus, in Figure 3.1, the level immediately below the ubiquitous expertises includes the groups labelled 'parents', 'car drivers' and 'office-workers'. These are very large groups, each made up of a considerable proportion of the members of society as a whole, but in representing them this way it is being implied that parenting, driving a car and office working are expertises acquired in secondary socialization and are associated with a spoken discourse that is not trivial to mimic when faced with a determined interrogator who already possesses the corresponding expertise.[3] In contrast, mere sets of people, such as those who wear brown laces in their shoes, don't have that kind of distinctive culture so don't count as social groups. As we go down the cascade of specialties, we also encounter scientists and, further down, gravitational wave (GW) scientists and if we had the space to draw them, we would encounter indefinitely more kinds of scientists, sub groups of scientists, and many other groups of professionals and people with special interests. The topic to be discussed here is the interaction between society as a whole, represented by the ubiquitous expertises, and the culture of the groups embedded within it, with the groups of scientists having special salience for the book as a whole and being distinguished from the 'profession' of science by the distinctions made in the descriptions of the seventh rows of Table 1.1.

It is because truth is so obsessively central to science that it has the potential to feed the value of truth back into society as a whole. But this has to be a re-thought science. Given what we know now about science as a result of over fifty years of science studies, and the influence of postmodernism on academic and public debate more generally, we must show not just how to disassemble scientific truth but to reassemble it again. Only when this is done will we be able to see how science might act as a role model for the bearing of moral truth on decision-making in general.

Postmodernism takes the meaning of words to be a matter of the interpretation of the reader not the intentions of the author. The ways of thinking which see science as a literary construct have been deployed in real-life politics: eroding the meaning of truth in the wider society leaves what counts as reality in the hands of politicians. This has been done quite deliberately in Russia (as referenced in note 10 in Chapter 1) and less calculatingly in America and Britain. So, it is important to understand that the critique of science as 'less than perfect' under Wave Two does not automatically lead to the major change in the role of science in society that some academics, exemplified by the democratizers, claim it must. The critique of science does *not* mean that the findings of science must fall into the domain of public choice, with all views being treated as equal, as might be reasonable in the interpretation of, say, works of art. With specific exceptions, scientific expertise, like legal expertise, is an elite expertise that is located in specific social groups and needs to be judged by peers. The restriction of the legitimate interpretation is justified by the expertise of the scientists and, above all, the formative aspiration of the institution – to find the truth and act truthfully.[4]

Science in society

Science, then, remains a crucial institution in democracies even though it is not the set of autonomous procedures standing above society and producing perfect knowledge that it was once thought to be. Science is imperfect and fallible, and this is especially noticeable when it feeds into the short-term demands of policymaking. But it is still an object lesson for how to how decide with integrity and still the way to bet when it comes to advice about the observable world. Seen this way, science does not replace politics, it helps to create the conditions for politics.

Many of the arguments presented here are directed at fellow academics and others concerned with the political interpretation of the meaning of science. But science and its relationship to truth must be understood more widely if democracies are to survive. The trouble with much traditional talk in the public domain about the importance of science is that it amounts to science popularization or science advertising, both drawing attention to the crown jewels. A veritocracy cannot start from a distortion of the truth.

Worse, starting from there is dangerous when it comes to policymaking. The public, if they look to scientists to answer policy questions after having been led to expect an instant kind of certainty, will either be misled by politicians, who assume that the public needs or wants such certainty, or disappointed when they discover it is not available. Widespread distrust of scientific experts is one possible reaction, though the COVID pandemic suggests that distrust of politicians is another possible outcome. In either case, the short-term horizon of policymaking is too short for science to reach a robust consensus on many new problems. We have our own slogan that covers this point: 'the speed of politics is faster than the speed of science'. The best we can expect of science when it is put under that kind of temporal pressure is that it continually *aspires* to deliver the truth even if it cannot guarantee it. That science *tries* to tell the truth, continually invents methods and procedures to help it tell the truth better, invites criticism, and is honest about what remains uncertain, is enough to make it an object lesson in acting with integrity. This is the best we can expect of science, but it is also what, in a veritocracy, we should expect of our politics.

Moral truth – the *determination* to try to tell the truth – is robust in the here and now even if success in accomplishing truth-telling is not always guaranteed. So, we know an uncorrupted science will tell the best kind of truth that can be achieved, even if that means saying that the truth about the observable world remains elusive at this time or that. That is how science and a truth-telling democracy – a veritocracy – support each other: science is an object lesson in acting with integrity and science is the best institution there is for providing answers concerning the observable world to feed into policy questions, not because it is always right but because its practitioners are trying harder to be right than anyone else and trying harder to find procedures that will minimize their almost inevitable biases. Absolute certainty about observable truth has to be found elsewhere – in magic, revelation, conspiracy and the like, institutions upon which we do not want to base our societies and policy choices, or so we hope.

Summary

This chapter has introduced the fractal model of society, and the distinction between ubiquitous and specialist expertises. This allows us to extend Durkheim's insights about the role of professional ethics in informing civic values to include a much wider range of social groups and show how to reverse the focus of much contemporary science studies by emphasizing the potential for the values of science to inform the wider society.

4

Defusing the tension between elite institutions, the public and society

Introduction

We have analysed different descriptions of science, but the classification of science does not stop with outsiders' accounts. Science itself, even when it accords with our description of its formative intentions, comes in different kinds. Disciplines differ from each other in terms of their content, methods and even the balance between theory and experiment. The previous chapters described science at a relatively high level of abstraction, identifying the shared values and aspirations that inform what it means to act 'scientifically'. But if we want to understand public participation in science, we have to explore the different ways of doing science, legitimate and not so legitimate.

Here we develop a new typology of science, inspired by the distinction between normal and revolutionary science but developing it further. The principal distinction is between sciences where the contextual assumptions are being continually contested and those where they are taken for granted. There is a very important interstitial category, novel to this analysis, where the contestation between founding assumptions is disguised in public discussion or disappears from view at the research front due to the reification of modelling techniques, so the science appears consensual when it is not; this interstitial category is manifested in what we call 'ubiquitous science'. The main three-way partition is then further sub divided, in the later part of the chapter, to identify the kinds of science where it will be useful or necessary to include experience-based expertise brought in from outside the scientific community. Though the entire drift of our argument is that both scientific research and experience-based expertises are the properties of elites, when it comes to upstream choices about which research to support, and downstream choices about how to exploit the outcomes, it is reasonable to include the wider public in the decision-making in virtue, not of their expertises, but their political rights and choices. The distinction between the research front on the one hand, and upstream and downstream on the other, is not always crystal clear but it has to be the starting point of the analysis.[1]

The main argument of the book is that science can support the maintenance of integrity in society as a whole. Figure 4.1 is an *aide memoire*. It reproduces the fractal model emphasizing the upward pointing arrow in Figure 3.2 linking the institution of science to the distinction in the ubiquitous domain between truth and lies.

The moral basis of science could provide the basis for a growth of trust within the wider society, but there should still be some occasional feedback

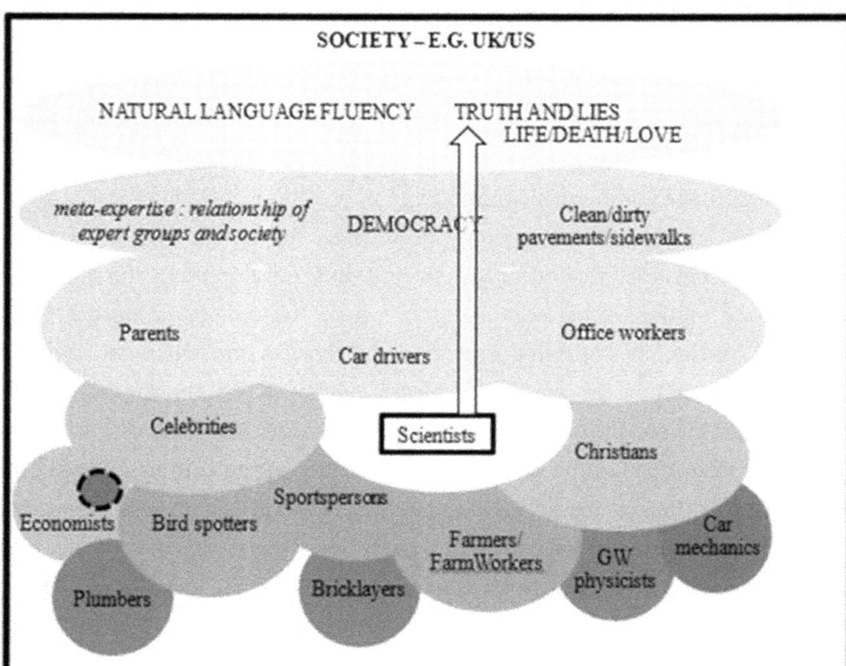

Figure 4.1 The fractal model of society emphasizing the role of science in veritocracy (created by Harry Collins).

from outside the institution of science under specified circumstances. The relationship between science and society is much more porous than we used to think but the boundaries that surround elite science must be preserved if science and expertise are to retain any meaning. We believe we can strike the right balance between strong boundaries and porousness by using the fractal model and considering the different domains of science that are found beneath the mid-level category of 'scientists' that is highlighted in Figure 4.1.

We now know that the entirety of knowledge relevant to science is not to be found in the specialist domains of the sciences. Some of that expert knowledge is found in specialist domains not normally thought of as belonging to science. As we have said, recognizing that there are relevant domains of experience does not mean the distinction between ubiquitous and specialist expertise disappears, as it does when experience-based expertise is described as 'lay expertise'.

We analyse a range of portals into science using the distinction between specialist and ubiquitous expertises as it bears on different kinds of science, the discussion being centred round Table 4.1: we will consider interactions at the research front of the science, upstream of the research front and downstream of the research front in Table 4.2. The more detailed typology of different types of science that we will work out provides the approach to the specification of the exceptions to the treatment of scientific expertise as found only in the profession of science.[2] Among the exceptions are the exercise of virtual diversity, which we will come to in Chapter 5.

Different kinds of science

A principal intervention in the classification of sciences was Thomas Kuhn's distinction between 'revolutionary science', where traditional founding assumptions are challenged, and 'normal science', typified by traditional ways of thinking and doing. The archetypal 'revolution' was the switch from Newtonian physics, with its conservation laws, to Einsteinian physics, in which mass and energy were interchangeable and the speed of light was an upper limit. As we explained in Chapter 2, Kuhn also pointed out that the textbook accounts of discovery were mostly wrong.

Kuhn also came up with the insightful term, 'the essential tension' (Kuhn 1959, 1977), which describes the desire in science to invent revolutionary ideas as in tension with the need to recognize established authority if science is to

Table 4.1 A classification of kinds of science with examples.

Salience of assumptions	Type of science	Description of science	Example science(s)
1	2	3	4
Assumptions a salient feature of day-to-day work in the science	**Fringe**	Try to overthrow not extend mainstream assumptions. Scientists marginalized though may work in universities	Electric universe theories
	Revolutionary	Critiques certain mainstream assumptions; retaining and extending mainstream if possible	General relativity and quantum physics at the turn of the twentieth century
	Uncredentialled Experience-based expertise	Experience of non-scientists relevant to mainstream in novel settings; their scientific socialization is problematic	Role of indigenous or local knowledge. Role of expert patients in medical research
	Most human sciences	Human actions can continually confound generalizations	Historical sciences/ macroeconomics. Effect of lockdowns
Assumptions should be salient but are ignored or suppressed	**Ubiquitous sciences**	Disagreement prematurely erased from public and political scientific debate.	Ubiquitous economics
	Reified models	Models mistaken for real world	Tendency of microeconomics/ string theory/ anthropomorphic AI/ maths worship
Assumptions rarely noticed in the day-to-day work	**Adventurous normal**	Theoretical foundations established but technical challenges unresolved or complexity confounds. First type can evolve into 'crown jewels'	Early Michelson-Morley/ early GW detection
	Normal	Mainstream puzzle-solving activity providing incremental addition to knowledge	Contemporary climate change. Particle physics

Salience of assumptions	Type of science	Description of science	Example science(s)
	Hyper-normal	Assumptions and methods are highly routinized, affording short apprenticeships, easy cooperation with industry and even automation. DIY is possible.	Molecular biology such as gene sequencing and editing
	Mythologized	Retrospective and idealized accounts of historically significant episodes in science (e.g. crown jewels)	Michelson-Morley experiment as represented in standard physics textbooks

proceed in an orderly and productive manner. Since Kuhn's work, a number of other variants of this kind of distinction have been described. Table 4.1 can be thought of as a development of the original Kuhnian insights.

In Table 4.1, column 1 provides the foundation of a new three-way classification of science. This classification is drawn from close observation of scientific practice and has to do with the extent to which the science is settled or sedimented. The main distinction turns on the extent to which the assumptions on which the science proceeds are a regular part of the deliberations in day-to-day scientific work. This is a central feature of the science in the upper part of the table but rarely or never a feature in the lower part of the table; in the interstitial category the assumptions are not part of day-to-day discussion, but ought to be.

The interstitial category

The top entry in the interstitial category is 'ubiquitous science', a kind of science where the public, the political domain and the elite of science research appear to merge. The public and political debate is stripped of the more profoundly contested aspects of the science and transformed into something more certain and taken-for-granted that can be presented as public property. This works through the distinction between inside and outside science described in Chapter 1 and represented in Figure 1.2. Ubiquitous sciences are not popular or popularized sciences: all parties acknowledge that a popular science is a watered-down entertainment or propaganda exercise, something like popular physics, with no sense that the public could actually grasp the technicalities well enough to make a judgement about whether a physics claim was true or not; popular science has about the same input to real scientific judgement as science fiction and everyone knows it if they think about it. In contrast, in the case of ubiquitous science, the public is invited to make a scientific-seeming judgement and to contribute to debates in the public and political sphere, usually to affirm the scientific findings as technically sound and act upon them.

One can see this use of science in description 6 in Table 1.1, and how the call for democratization (Chapter 2) fits with it. One can also see it at play in the MMR revolt, where a proportion of the public considered that their observations contributed technically to the decision about the safety of the vaccine.

In the case of MMR, politicians were mostly not encouraging the public to make these judgements (though some Conservative politicians in the UK were heard to make positive noises about public choice), but the democratizers coming out of the co-production idea in science and technology studies were encouraging them. The subsequent exposure of the unethical and financially self-interested nature of the actions of Andrew Wakefield that triggered the revolt had, unsurprisingly, little impact on the views of the public or, more surprisingly, on the views of the democratizers.

Another more sinister example of ubiquitous science is the eugenic science of the 1930s as exploited in Nazi Germany. The notions of genetically identifiable uber- and untermenschen readily became public property and supported the National Socialist political view of the world. It may be that the same analysis would have applied to Lysonkism in the Soviet Union, but we do not know the extent to which the debate about the inheritability of acquired characteristics was carried out in public.

In Chapter 6 we will describe 'ubiquitous economics' in these terms – a science which is discussed in newspapers and political forums as though it was a ubiquitous expertise belonging to the public even while the assumptions and conclusions are still heavily disputed among sectors of the expert academic community. It is these kinds of sciences that populate category 6 of Table 1.1. These, to repeat, are sciences where the public have been persuaded that they understand sufficiently well to join the discussion – distance lending enchantment – whereas the crucial debates are far from settled within the scientific establishment itself (see also Figure 1.2). For the sophisticated politician, or at least for political actors with sophisticated advisors, it is category 5 – *Just another literary construct* – or some watered-down version of it, that can provide the conceptual invitation to invite a ubiquitous science into the political sphere.

The second category of interstitial science is 'reified models'. Many sciences use models, nowadays mostly computerized mathematical models, as ways of theorizing and thinking about the world. It is when the models take over from the worlds that they represent that this kind of interstitial category of science emerges. The construction of models is a technically demanding expertise, and that amount of effort put into to building a beautiful 'machine', either a physical machine (as with early hydraulic models of the economy), or the hugely more intricate mathematical models, can provide the illusion that science is being done whereas really it is an artificial world that is being constructed. We'll see this kind of interstitial category also appearing in both micro- and macroeconomics.

Column 2 of Table 4.1 names the types of science that fall into these categories, while the third column provides a brief description of each, and the fourth some examples of specific sciences. We'll now work through the table from the bottom upwards, starting with the sciences for which the basic assumptions are not actively questioned.

The bottom row is more a distorted representation of science than a way of doing science. The crown jewels of science are actual episodes of science, typically characterized by their contribution to crucial discoveries or theoretical insights, but what makes them mythical is the way their re-telling strips out all uncertainty and doubt, leaving only the clear and conclusive result. The Michelson-Morley experiment is a classic example of this: it was not intended to 'prove' General Relativity but is now routinely described as having done so. In fact, as discussed in Chapter 2, it is only in retrospect that the null result becomes the expected result and the experiment is, from that point on, hailed as the first demonstration that the speed of light is constant.

Moving up the table, we enter the realms of scientific practice in which the core assumptions are all taken for granted. These next three types of science are all variations on Kuhn's normal science. Hyper-normal science is a newly identified type of science that has been exemplified by molecular biology.[3] In hyper-normal science the founding assumptions of the field are as established as can be, as are its research methods and approach. The methods and procedures of hyper-normal science are so well established as to be beyond question by the time a PhD-level apprenticeship has been completed. This means that anyone with a PhD-level training can work confidently with others located remotely, knowing that conceptual confusions or problems of tacit knowledge will not be an important feature; this kind of science is often ripe for computerization and even automation because the models are so sedimented. One consequence of procedures and materials becoming ever more standardized is that it becomes increasingly feasible for the same techniques to be used by those outside the scientific community, there being no seam between the science and its contributing industries but also making synthetic biology done in the home, using off-the-shelf purchases of equipment and materials via the internet, a possibility – flatpack science, as it were – raising concerns about ethics and biosecurity.

Moving up the table, the next row contains what we can think of as regular, less extreme, normal science. Here the possibility of questioning the field's assumptions does not arise but tacit understandings and practices are an ever-present feature, and research work involves developing new techniques and instruments collaboratively. The particle accelerators at CERN provide a good example of this kind of normal science: the standard model and the fundamental particles it postulates form the basis of the research programme, but larger and more complex machines are continually needed to detect ever more elusive particles. As each new experiment requires new detectors, each experiment represents the cutting edge of particle physics at that time. The other end of the cost spectrum is represented by the TEA-laser study.[4] Research of this kind is a continuing apprenticeship, and the required skills are not accessible to those outside the field.

The next category is 'adventurous normal' science, where the core theory is not in question but a radical change in experimental technique or sensitivity is involved. The field of gravitational wave detection is a good example of an adventurous normal science: the physics on which it depends was not in serious question at the time it started to receive major NSF funding (roughly the early 1980s) but the proposed measurements were close to the bounds of credibility. After several decades, gravitational waves were first detected using laser interferometers with arms 4 km long sensitive to length changes down to less than a 1,000th of a diameter of a proton and recognized with Nobel Prizes in 2017 but with the assumptions upon which the first discovery was based (such as the speed of transmission of the waves – Chapter 2) still marginally present in scientists' awareness. The detection of gravitational waves is now becoming 'regular normal' rather than adventurous as the new field of gravitational wave astronomy develops with hundreds of detections to be analysed. Nowadays, the science of climate change is in a similar position.

The interstitial category has already been discussed in the introduction above in Table 4.1. We can add string theory as an example, where mathematical modelling seems to have taken over from the world, and large parts of artificial intelligence, where the capability of the computer programs is described in anthropomorphic terms, as though the deep problem of human intelligence had already been solved just by mimicking some of its aspects. Maths worship is self-explanatory: too much respect for the role of mathematics in our understanding of the world can easily lead to the displacement of science by mathematical exercises. Chapter 6 of this book will discuss economics as exemplifying both kinds of interstitial category science. The reification of model building based on a stripped-down version of a human – *homo economicus* – as a set of preferences can easily take over from the more cautious teaching in universities where all the complexities are reintroduced after the initial idea has been presented. We will argue that economics' involvement in politics and public debate is what has given rise to this premature acceptance of a certain set of assumptions, especially in ubiquitous economics. The same might be said of the Marxist version of economics: insofar as Marxists cleave to the idea that their account of the economic and political evolution is 'scientific', even though it is part of the understanding of the relevant politically leaning public at large. In today's world, Marx-based ubiquitous economic science is a far less successful than the one turning on the importance of markets. One thing that will be revealed about this kind of science is that it exemplifies the overemphasis on the left panel of Figure 3.2 – the influence of society as a whole bearing on the science and distorting its search for correspondence truth.

Moving to the upper category in column 1, we encounter, at the bottom of that category, a first type of science where the working assumptions are a feature of day-to-day work. This pretty well covers all the social sciences and some of the human science such as psychology. 'Historicism' is the name given by Karl Popper to sciences of history because they are mostly single series of events based on assumptions about what can be observed in one or two societies and neither replicable nor falsifiable from observation and hence highly assumption bound. Though there are a series of contributory factors, varying assumptions probably contribute to what has become known as the 'replication crisis' in psychology and medicine. One crucial assumption is that the 2-sigma level of statistical significance merits the publication of a finding whereas what amounts to the long-term if unintended empirical observation in physics, of the reliability of different levels of statistical significance, indicates that 5-sigma is nearer to the level needed for reliability (probably because of the swamping of the unknown effects of systematic errors which are, to make matters worse, much more severe in the social sciences).[5] Even in the more exact and testable of the social sciences, for instance, macroeconomic modelling of economies, it is well understood that the predictions that emerge from the models will vary hugely according to the assumptions deployed at the input end. It is the hiding of such assumptions that give rise to the interstitial category. What one might call 'lockdown science', which is the attempt to make policy to counter the recent pandemic based on the predicted effect of lockdowns on social interaction, is part human science, part epidemiology, part economics of health care, part child psychology and part moral judgement: it is quintessentially assumption based.

The category of 'Uncredentialled Experience-based expertise' includes a variety of different examples united by the fact that the science in question is being applied in some new context and that additional work and expertise are needed to bridge the gap between the science and the context.[6]

Returning to the kinds of science that we might find within each type, the least controversial of the 'contested application' type is probably medical research where the potential for expert-patients and other treatment activist groups to contribute to the research is well-established within the social science literature (Arksey 1998; Rabeharisoa and Callon 2004) and, in some cases, recognized by the medical profession itself (Epstein 1996). That said, there are also examples of medical experts being resistant or reluctant to accept patient experience as expertise (Callard and Perego 2021; Eyal 2013).

Moving 'upward' within this category, the next kinds turn on the local circumstances and local experience which cannot be known to a generalizing science. The template for these cases is the application of a scientific theory or method to a local setting that requires some context-specific adjustments to be made. As the accredited scientists do not, in general, have the necessary knowledge, experience-based expertise from the relevant local community is needed. The sheep farmer study that we have mentioned is a standard example of this genre, with the government scientists needing (but not necessarily recognizing that they needed) the expertise of the sheep farmers in order to refine various assumptions about the behaviour of sheep, their husbandry and the response of the land to radioactive fallout. There are many other examples, however, as similar concerns will arise whenever 'standardised' policies that fail to reflect local variations and practices are imposed (see, e.g. Wickson and Wynne 2012). Another example is indigenous science, but we will deal with this more fully in the next chapter. In Chapter 5 we will try to resolve any problem to do with indigenous science's incompatibility with the idea of uniformity in a new way, by treating indigenous knowledge as a form of experience-based expertise and emphasizing the role of 'virtual diversity' in handling both. The integration of experience-based expertise into elite science is a kind of interdisciplinarity, with all its problems, but with the additional problem that the experience-based experts may not have been socialized into the norms and procedures of science as a whole.

A case of the final kind belonging to this category of science has already been mentioned in the 'human sciences' category – lockdown science. It's no surprise that there can be ambiguity about where human sciences end and where experience-based expertise begins, since the assumptions under debate often belong to citizens, not scientists. When a new threat emerges such as the recent COVID pandemic, medical science is needed to characterize the new virus so as to understand its effects on humans and how it spreads between them, but new public health knowledge about how populations will respond to interventions ranging from vaccines and mask wearing to lockdowns is also needed, with uncertainty about how different social groups would respond being as much a part of the policy problem as the basic virology. Wherever the human sciences are involved there are always likely to be salient but unresolved assumptions.

Finally, the same ambiguity of location applies to macroeconomic modelling which turns on formalized model of human behaviours and is explicit about resting on salient assumptions in its model-building practices. And, since economic policies are always changing, as is the wider economic environment, the world that the models attempt to represent is always in some degree of flux. Nevertheless, the need for forecasts and policy-advice means that macroeconomics has to act as if conclusions can be drawn despite the uncertainties and limited consensus which is typical of human sciences. The difference is that the situation is a chronic problem for macroeconomics, whereas for other domains it is usually resolved over time. Despite this the uncertainty associated with macroeconomic forecasts is rarely discussed in public, with governments often projecting economic performance several years into the future and exhibiting all the problems of a ubiquitous science.

The two uppermost kinds of science in this category are revolutionary science and fringe science. Both question the basic assumptions of established science in a radical way. It is tempting to say that fringe science is simply revolutionary science that has not yet given rise to a revolution but there is less continuity than this would imply. Fringe sciences have a particular character which can be described, in part, using Kuhn's 'essential tension' idea. Fringe scientists often appear to have an extreme focus on novelty at the expense of continuity. The balance between authority and novelty is distorted as compared to even revolutionary science, where scientists want to retain as much existing knowledge as is compatible with the rejection of certain basic assumptions. Fringe scientists, as at least one fieldwork-based study has indicated, want rupture not reform – they are not the 'reluctant revolutionaries' who often lead radical change *within* a science.[7]

Public portals upstream, downstream and at the research front

The way the boundary between elite experts and the public works depends on the point at which the groups make their encounter: upstream of the research front, at the research front or downstream of the research front. The public encounters science *upstream* of the research front where choices are expressed about priorities and preferences. These can include high-level questions about how big the budget for science should be and how that budget should be distributed among different scientific projects or more specific questions about how a scientific problem is 'framed' within the context of a wider political decision. For example, is the problem of how to dispose of North Sea oil rigs at the end of their lives a matter of keeping the oceans pure in some abstract sense, in which case there is not really a scientific question to answer, or something that depends, at least in part, on technical advice about the likely cost and efficacy of the measures needed to dispose of them in more or less environmentally friendly ways or the possibility that dumping them at sea might benefit fish populations by providing new habitats secure from nets? In such cases different stakeholders might have different legitimate interests. This kind of choice is part of the normal political process, so the public are entitled to be involved, even if the esoteric nature of the issues sometimes makes things difficult and problem can bleed into the research front with strong opinions on the political issues causing difficulties. Another example is lockdown science, where the way the problem is framed – is it just medical or should economic and future health policy be counted as part of the 'scientific' problem – influences the choice of expert advisors, in turn affecting the conclusions at the research front. This is a boundary problem which applies in certain cases but concentrating on these alone, as the democratizers tend to, takes attention away from the philosophical problem of reconciling societal values with universalism and uniformity. To repeat, the place to start is with clear distinctions, treating places where the boundary is fuzzy between the research front and upstream/downstream as exceptions.[8]

The way public choice in these matters of preference and priority outside of the research front is expressed might be indirect: by choosing between governments with manifestos that favour more or less resources going to science, or this and that particular science, or even more indirectly by watching TV programmes about astronomy, space exploration, cosmology or natural history, which is then taken as indicating that they want more of it. Or the choice might be expressed directly by supporting campaigning organizations or taking part in protest marches and demonstrations, thereby explicitly positioning themselves against research on nuclear power and in favour of more green energy research for example, or in favour of certain types of medical research but against research on genetic manipulation of crops (or vice versa), and so on.

To repeat, one important way in which this analysis differs from that provided by those who typically argue for the democratization of science is that only in special circumstances do we see these kinds of activities as contributing to the science; usually they are part of the political process that comprises science's social context. In contrast, because the democratizers do not see any boundary between science and politics, they do not – cannot even – recognize this distinction. It is, however, relatively easy to understand: upstream engagement involves attempting to influence the questions that are asked of science, the way the burden of proof around those questions is set and, in the extreme cases, it may even prevent the research from taking place at all. Usually, the lobbying and campaigning do not determine the outcome of the research and the citizens doing the campaigning and lobbying do not take part in the science. As we will show later, there are cases where some citizens – namely those with relevant experience-based expertise – should contribute to the science at the research front, but these are the exception, not the rule and even more exceptionally, highly activist citizens can come to acquire the scientific expertise needed to contribute directly – but just being a stakeholder is something different.

Downstream of the research front citizens engage with the findings of science that has been completed and disseminated. This might feed back into subsequent upstream engagement of the type mentioned above: enthusiasm for a type of science such as cosmology or natural history, as indicated by TV viewing, might help ensure future funding for this kind of research. Alternatively, the outcome of an inquiry or policy decision may be accepted or rejected, with the latter likely to lead to a new round of upstream engagement. In addition, citizens will have to decide how to act on those aspects of science that impact directly on their own lives. This will include things like where and how they seek medical advice, decisions about what they eat, choices about energy use and transport, and the like.

As long as we can see that upstream and downstream engagement differ from work at the research front, the public's contributions to the political and financial life of science do not cause the kind of philosophical tensions that we are exploring here. For example, there is no tension between upstream and downstream engagement and uniformity and universality – political interests in how much science is done and what science is done are not expected to be culturally uniform or exhibit universalism. It is possible that more enthusiasm for science might encourage respect for its check-and-balance role, but this would not change the nature of science. In a similar way, whilst the rejection of vaccination, masking against COVID and the endorsement of alternative cures have all been used by conspiracy theorists and populists in their attack on truth in general, there is nothing puzzling going on and no reason why science should be done differently as a result.[9]

The general public can, of course, contribute data to the research front by reporting changes in bird and insect populations, monitoring levels of pollution and contamination, or watching the skies for unusual events. They can also donate other resources, beyond their unpaid labour, by providing downtime on their home computers to analyse data streams. These interventions are possible because the skills or equipment required are widely available outside the scientific community. Whilst some, minimal, training and guidance is required, contributing to the research can be done with little more than basic observation and teamworking skills supported by the norm of honesty: in other words, by utilizing expertises that are ubiquitous or, at least, very high up in the fractal model.

Upstream → Research Front → Downstream: Interventions

Table 4.2 is an elaboration of Table 4.1 that summarizes and organizes these issues around the different types of science discussed in the previous section. Table 4.2 reproduces columns 1 and 2 of Table 4.1, and then adds three new

Table 4.2 A classification of kinds of science showing kinds of public intervention.

Salience of Assumptions	Type of Science	Public Input Upstream	Research Front	Downstream
Assumptions a salient feature of day-to-day work	**Fringe**	Conspiracy theories challenge elite institutions, including science	Autodidacts can contribute in-principle but rarely do so in practice	Adoption by conspiracy theorists using social media
	Revolutionary	Awareness of potential for revolution	No contribution to science	Awareness of major research events and results.
	Uncreden-tialled Expe-rience-based expertise	Lobbying and funding as extra-vigilant stakeholders	Experience-based expertise can con-tribute to science	Support or protest based on outcome
	Most human sciences	Accept/resist constraints on action	Citizen's actions are the data	Accept/resist constraints on action
Assumptions should be salient but are ignored or sup-pressed	**Ubiquitous sciences Reified models**	Public support encouraged by media campaigns	Public act as though they understand eco-nomics, taxation, enterprise, etc.	Public support encouraged by media cam-paigns
Assumptions rarely noticed	**Adventurous normal**	Encourage fund-ing	No contribution to science	Consume and enjoy
	Normal	Encourage fund-ing	May have expertise needed to contrib-ute to data collec-tion activities	Consume, enjoy, trust
	Hyper-normal	Encourage fund-ing	No contribution to science	Consume, enjoy, trust
	Mythical	Encourage fund-ing	No contribution to science	Consume and enjoy

columns indicating, respectively, how the public might interact with science upstream, at the research front and downstream.

As intimated, the research front is the crucial kind of interaction in the context of this book because it is where the epistemological friction is found. Here we describe the contents of these three new columns, following the analysis provided above: that upstream and downstream public interventions are mostly little different to interventions in any other political domain whereas interventions at the research front can threaten the whole idea of science and its roles as check-and-balance and object lesson in democratic societies: in sum, it is subject to the problem of extension and the consequences of failing to solve it.

The top row of Table 4.2 concerns fringe science, by which we mean science where the quest for novelty and rupture from the mainstream has created a cognitive activity too divorced from the mainstream ever to join up with it again. Its practitioners may believe they are pursuing correspondence truth but the way they are doing it takes them into a strange culture incompatible with uniformity and universalism: after all, witch diviners might also believe they are pursuing correspondence truth! In the case of fringe physics, for example, the field and its participants are characterized by an almost pathological desire for novelty that makes it difficult for them to reach consensus even amongst themselves.[10] At the research front, a significant proportion of those advocating these views are credentialled scientists working in universities but they are generally isolated within those institutions, unable to publish in mainstream journals and are uniformly judged by their departmental colleagues to be outdated at best.

In principle, given the emphasis on novelty, and the absence of formal institutional structures to control membership, it might be thought easier for citizens to join the fringe science community and work at its research front by teaching themselves the necessary content than is the case for mainstream science. In practice, however, this rarely seems to happen, especially in fringe physics where the dominant demographic appears to be retired engineers, though one might argue that they are half-way between credentialled physicists and outsiders. As the fringe science moves way from direct engagement with mainstream physics as, for example, in Erik von Danniken's popular books describing an alternative history of the world, the potential for lay citizens to contribute to the research front increases but the most likely role for the general public when it comes to the fringe sciences, particularly those that concern health and medicine, is as onlookers or consumers rather than contributors.

Fringe sciences also play a role in the wider debates around trust in science and the nature of truth. Conspiracy theorists and other sceptics will find anything that questions the existing order of authority appealing and, where authority includes elite science, fringe sciences provide a useful resource in these endeavours. There is no difference here with the exploitation of any other conspiracy, however, so the public support for fringe science poses no special epistemological puzzle.[11]

The next row down, revolutionary science, differs from fringe science in that those in the fringe are trying to reverse a revolution that has long since become the orthodoxy, whereas revolutionary science is a revolution in formation. Participants at the research front are typically located in mainstream research institutions and publish in mainstream journals, so there is little prospect of participation in this work by the general public. Indeed, it is unlikely that the public will know anything much about what is going on as these proto-revolutions are generally hidden from public view unless the mass media and, nowadays, social media find something about the potential revolution interesting. If this happens, then it is possible that the general public might become aware of the debates, even if they remain distant spectators in terms of any contact with the research front. For example, it is said that Eddington's eclipse observations, taken to demonstrate the truth of Einstein's theories, were widely publicized with a view to furthering peaceful relations between Britain and Germany following the First World War.

The third row is the 'Uncredentialled experience-based expertise' category and is where many of the case studies favouring 'public participation' as a consequence of indigenous knowledge, local expertise and lived experience of health conditions are found. The 'localness' might be in terms of physical location, such as in the case of environmental controversies involving indigenous communities or people who live close to any industrial installation that causes them to be concerned about their health or quality of life. Alternatively, the 'local' nature of the knowledge might refer to a narrowly bounded group of the public such as those afflicted with rare illnesses, or those with particular skills such as farming. Either way, it is this localness that makes the research front column salient, and the language used to describe these potential participants so important. In much of the published literature, these groups are described with adjectives such as 'lay' or 'ordinary' in order to stress their separation from the institution of science that is, in most cases, aligned with other social elites and part of the complex of interests and institutions that is ignoring or harming these groups. This emphasis is dangerously misleading. What makes these groups so salient is that their experiences give them a unique expertise that is directly relevant to the controversy at hand. This uniqueness might be living as a Maori, living as a sheep-farmer, living as a farmworker who sprays weedkiller, living next to a plant that emits pollution or contamination or living with a chronic illness. Whatever the experience, so long as it is not ubiquitous, it gives these uncredentialled experts a special contributory relationship with the corresponding science because scientists' regular socialization is not likely to have included experience of these physical or conceptual localities.

The general public do not have this esoteric socialization and cannot contribute for the same reason, so to imagine that some variant of democratization of science will resolve scientists' local socialization deficiencies is a fallacy based on wishful thinking at best. In the next chapter we will try to establish this point in a more formal way.

This does not prevent the wider public from taking a side in such a controversy and seeking to influence its outcome through other channels. Upstream we may expect lobbying for more attention to be paid to the local interests and conditions, with the same interests feeding back from downstream. Interestingly, the local communities enrolled as experience-based experts can, and do, engage in all these activities as well. Indeed, they are the key stakeholders in many respects as it is their community or livelihood that is under threat. This means they are likely to be especially vigilant and sensitive to subtle issues of framing in upstream discussions and quick to spot any 'spin' put on the dissemination of results. On the other hand, whether this personal stake means their views should get any special or extra weight is one of the standard problems of political decision-making.

By recognizing that the general public can contribute only to upstream and downstream activities, whilst experience-based expertise can contribute to the research front too, we can see how interests and experience can be separated. Neither the general public nor most stakeholders have the experience or expertise needed to contribute at the research front and so they cannot have the same rights in this domain as those who do. In contrast, where social groups do have relevant expertise and experience, their relationship with the science is not only different but, precisely because it is a specialist expertise, it is incompatible with the notion that expert comment is a democratic right. As we will see in the next chapter, managing the integration of expertises is not straightforward but the differences in status, resources and power mean that the absorption of the experience-based expertise should, in almost all cases, be managed by the scientific community.

Moving to the next level we have the human sciences. One recent example mentioned was the lockdowns that took place in attempt to reduce the spread of the COVID pandemic. For both upstream and downstream engagement, we find strong public interest in the epidemic and its treatments, with a strong tension, most obviously in the United States, between support for science and policies that act on it, and sceptical claims such as that masks or lockdown measures were unnecessary. Mass media and social media coverage meant that citizens were aware of these competing issues but the balance between them varied across different countries, with the United States under President Trump being notably more hostile to preventative measures than other established democracies. In the United States, we find public interest in the epidemic and its treatments being used as political symbols, notably with Trump claiming that masking was unnecessary and, led by the President, those attending his political rallies refusing to wear masks. Trump also proclaimed that alternative treatments would cure the disease, embodying the legitimacy of interventions by the unqualified at the research front – something that illustrates his desire to demonstrate the political substance and location of what counts as truth. In a nightmare world, Trump's interventions would lead to another ubiquitous science with politicians and the public deciding on the proper medical treatment of pandemic outbreaks just as they currently decide on the appropriate levels of taxation and government intervention in the economy.

In other democracies we also find, if not widespread medical debates (though there are always fringe movements), widespread economic debates related to the pandemic, since lockdowns are economically damaging. The rhetoric sometimes deployed was that COVID-19 was no worse than influenza and should be treated as of no more importance, allowing the economy to function much as before. It is not hard to see why a proportion of the public would lobby against epidemic science of most sorts and why many would react negatively to medical advice with the usual feedback from downstream to upstream. It is also easy to see how upstream and downstream interests would encourage the abandonment, among the public, of the usual authoritative role of scientific expertise, the public effectively invading the research front. It is hard, however, to find any legitimate feedback from the public at the research front though there were misplaced anecdotes about individual's exposure not leading to infection, combined with conspiracy theories concerning the untrustworthiness of drug companies and equipment suppliers, with stories of death or illness following vaccination such as also as drove the MMR vaccine revolt.[12]

The only way public understanding might legitimately bear upon this topic is in the field of public health, where some sense of the extent to which ordinary people would obey lockdown rules and the extent to which vaccines would be taken up was needed. The public might have more sense of this than the medical scientists, but the knowledge would surely need to be backed up with some survey data which would treat the public as respondents rather than experts.[13] Individuals who downloaded apps and symptom checkers contributed to the developing understanding of science in much the same way that participants in annual bird counts contribute to research in ornithology and, over time, expert patients play an increasingly significant role in the characterization and treatment of 'long COVID'.

COVID is just one extreme example of the relationship of the social and human sciences to the public – the final category in the upper portion of the table. Another example is macroeconomics. As mentioned above, this will feature prominently in Chapter 6. Given the impact of economic policy on almost every aspect of daily life, it would not be surprising to find both upstream and downstream engagement with the science that underpins it. Upstream and downstream, citizens act, albeit mostly indirectly, through their support for organizations that fund either economic research or seek to influence its research agenda. Likewise, the contribution to the research front is also indirect, in the sense that it is their economic behaviour – how far will they travel to seek work, how much they save and how much they spend – that is the very stuff of economic modelling. As such, it is not surprising to learn that there are a whole range of formal indicators that provide data on all kinds of sentiment and intentions. In addition, and although this influence is greatest when it comes to the use of the models for forecasting, there is more in-person input from social networks and contacts that is used by forecast teams to fine-tune the outputs.[14] Of course, all these judgements and adjustments contribute to the uncertainty surrounding the forecast but, as noted earlier and to be discussed in more detail in Chapter 6, little, if any, of this makes it into the ubiquitous economics that is the basis of much political discourse and which retains its simplified and distorted enchantment.

Macroeconomics can stand for the social and human sciences in general because wherever they bear upon policy, we would expect citizens to feel they have something to say about them both upstream and downstream. And, of course, it is human actions that constitute the substance which the human sciences investigate, so citizens' contribution to the research front is central, even more so when the method of investigation is participatory.

Continuing down the table we reach the interstitial category, where assumptions are routinely ignored, prematurely in terms of the stage of development of the science. Major political parties in many Western democracies base their campaigns on the invasion of culture by ubiquitous economics and, politics being downstream of culture, elections reinforce the feedback cycle. In turn, we will suggest in Chapter 6, that there is another feedback cycle into the research front of economics as it is found in universities. But this is an exception that proves the rule since it is a malign kind of feedback, supporting the *profession* of economics but weakening rather than strengthening economics *as a science with correspondence truth as its aspiration.*

In the lower part of the table, the first two types of normal science are very similar when it comes to their relationship with the wider society. Here both the upstream and downstream contributions are linked to the funding of particular kinds of science and the framing of the research problems it is tasked with addressing. As for the research front, there is very little that the public can contribute to either of these kinds of science, except in those case where relatively ubiquitous skills or technology allow some aspects of science to be outsourced to citizen scientists. In most cases, however, the specialist expertise and equipment required mean that any contribution to the research front will be beyond the capacity and resources of non-specialists.

Hyper-normal science may be exceptional when it comes to the research front. Here the availability of off-the-shelf technology and materials mean that the domestic pursuit of molecular or synthetic biology is a real possibility. In this case, what makes the elite institution of science different is not its access to technology or knowledge, but the shared value commitments and related review processes through which scientists choose to regulate their work. In contrast, the realm of DIY synthetic biology has no such commitments or oversight, raising concerns about the terrorist potential of the amateurs creating novel infectious viruses.[15]

Finally, there is the mythical science exemplified by the stories about science's crown jewels. Given that these are no longer active fields of research, there is no possibility of either upstream influence or participation at the research front. What there is, however, is plenty of opportunity for downstream engagement through the consumption of popular science programming and publications and, for most people, the school science curriculum in which such reified accounts of science feature prominently.

Summary

Table 4.1 develops a new categorization of types of science with a particularly important middle category. Table 4.2 is a case-by-case analysis of interventions into science by members of the public based on Table 4.1. The problem is different depending on the phase of the science that is being considered – upstream, research front or downstream – with the research front exhibiting the crucial philosophical tensions. At the research front participation by members of the public is legitimate only in special circumstances. One type is exemplified by outsourcing data collection that rests on ubiquitous expertises to citizen scientists. Bird-count surveys are an example with a more troubling variant being DIY molecular biology where automation renders the expertises available to the citizen.

The second type is experience-based, expertise where legitimate participants at the research front are experts in some relevant domain but without scientific credentials or, more importantly, scientific socialization. In the next chapter we'll provide a solution to the more important problem.

5

Virtual diversity and universalism

Introduction

The previous chapter looked at different kinds of science and at the places where members of the public could make contributions, notably in the matter of distribution of resources, and it also looked at where uncredentialed but experience-based experts could contribute at the research front. Where the expertises required to contribute to a science are ubiquitous, or near ubiquitous, all citizens can contribute to all aspects of the science, for good or ill. The problem dealt with in this chapter is how to incorporate elite experience-based expertise into science.

Experience-based expert citizens sometimes represent distinct social groups. This has encouraged some philosophers to argue for the standpoints of more distinct social groups to be represented in the scientific profession – to argue for more diversity in science. This is an attractive idea because at first sight it looks as though experience-based but currently uncredentialed experts would automatically be included in the scientific profession, solving the problem of under-representation of all kinds of groups at the outset.

In this chapter, we argue that this 'demographic solution' will not work but that another solution is available. We argue that where epistemological considerations are at the forefront the required diversity is always context-specific and that the development and sharing of interactional expertise to create what we call 'virtual diversity' provides a more realistic way of enabling expertise

We are grateful to the editors of *Social Studies of Science* for permission to publish this chapter because it follows closely the arguments and much of the text of Collins, Evans and Reyes Galindo 2024. Actually, this chapter was written first, but the contingencies of publishing meant that the SSS version was *published* first. Readers might be interested to learn that the journal publication is followed by four critical comments and a response from us.

(Collins, Evans and Reyes Galindo 2025)

to be shared than full participation by every relevant group. More diversity in the profession of science is desirable purely on grounds of social justice and it does not need to be justified in terms of epistemic goals. In any case, as we will show, epistemic goals can only rarely be achieved by these means.

On the other hand, we now know that in spite of the aspirations and the procedural safeguards there are, *in fact*, cultural variations within the sciences, giving weight to the standpoint argument. For some, as has been seen in Chapter 2, it has been a short step to the idea that the sciences are little different to any other cultural institution in terms of their embedding in societal influences so that societal influence should be embraced in science rather than abhorred, and uniformity and universalism abandoned.[1]

Before going on we need to eliminate one possible misunderstanding of what is meant by uniformity. The aspiration to institutional uniformity across science does not mean that the related policy decisions have to be the same everywhere. What scientific uniformity across cultures means is that scientists should aspire to agree about such differences as there are, such as the varying features of different settings which precipitate different policies in different settings even if the science is the same everywhere. If, deep down, nature is uniform, then a science of nature should be same everywhere including the way it responds to environmental and cultural variations. That is why there can be a science of society even though we know from the outset that, quintessentially, societies differ from place to place – it is the analysis of the differences that is the foundation of consensual social science. The thing that is uniform is the understanding of how things, including social things, work. Given this, the human variability and frailty that might work against the uniformity of perception are subject to *procedural safeguards* as exemplified the rightmost column of Table 1.2.

The tension between the cultural influences on science that cannot quite be eliminated by procedural means and the aspiration to uniformity is one thing we have to deal with here. Whereas the early debates about cultural variation and influence took place in a context where the authority of science could be assumed, we are now dealing with it in the context of the new fragility of science as a check and balance given the rise of populism and the potential slide into fascism, most graphically illustrated by the erosion of truth under would-be dictators such as President Trump. The challenge is to find a way of handling today's much greater awareness of science's permeability to cultural variation, to local interests, to individual expectancy effects and the pernicious effects of the internet and social media, without abandoning science as a vital institution in democracy.

The 'demographic solution' and its failure

To go back to the standpoint response, it is to demand that science itself absorbs and ameliorates any lack of uniformity across cultural differences, strengthening the epistemic warrant by embracing a greater range of criticism from a variety of standpoints: in short, there should be more *diversity* in the scientific community, and this would give rise to 'strong objectivity'. Groups who are currently outside the scientific elite should participate more fully in the scientific research process, resolving the political suspicions, reducing the epistemic weakness if there is one, and certainly ameliorating the social injustice of the exclusion. Outside groups who it has been said should be brought into science from the epistemic cold include workers with specialist skills; sufferers from certain medical ailments who have expertise relevant to the related science as a result of their experience; indigenous groups who have relevant specialist local knowledge and local interests; local groups in Western societies who have special knowledge of local conditions near industrial plants and the like; women and other disadvantaged groups, who are underrepresented in most scientific research yet have specialist contributions to make; and members of society in general who would make science more democratic and could also make sure that scientific research is done in accord with general societal values. Taken together these appeals represent a move to blend science with the citizenry which is more nuanced and exact than that of those we have referred to as the democratizers but still risks abandoning uniformity and universalism.

We think bringing outsiders into science can work without threatening to erode science as an institution, but as shown in Table 4.2, we think those occasions need to be understood case-by-case, not as a general prescription. We can see the difference between the two approaches by considering the following statements which, as it happens, relates to the way the case might be made in respect to the indigenous knowledge of the Maori people of New Zealand:

> Maori development research must be by Maori, for Maori, and work from a Maori worldview.[2]

The alternative being argued for here is along the lines:

> Scientists agree that the Maori have certain special needs and local knowledge that should be respected and, where appropriate, utilised in any attempt to apply 'this or that scientific principle' to their situation.

Comparing the two indented phrases, the first sentiment puts the local people in the driving seat of a local science, the second keeps the scientists in control of a uniform and universalistic science which recognizes differences. To repeat, the fact this illustration has to do with Maori in particular is not important: the same indented phrases could be applied to differences between cultures such as Mexican, Christian, gender as highlighted by feminism, and so forth.

Comparing the two, we can see that the first imposes no hierarchy on the different kinds of science; they are equal but different and each group is free to believe what it wants and act accordingly. In contrast, the second assumes that there is a cross-cultural institution called science that has some responsibility for listening to and somehow integrating the different perspectives. From the perspective of political liberalism, the first seems morally preferable as it respects the different perspectives and does not try to constrain the ability of people to live in the way they choose; in contrast, the second seems less liberal, and more authoritarian, in that it implies that some determination will eventually be made as to the 'right' or 'best' way to understand an event or question. To help avoid any too quick assumption that the first approach must always be best, remember that there are other institutions, such as international sport and international justice that aim, like science, to work towards uniformity and universalism rather than diversity and such institutions are usually approved of by citizens from many kinds of society.

In today's political context, where universal values are being attacked by nationalistic populists, it is the second approach that needs to be defended. Even in the context of political liberalism, individual freedom to act is limited by the duty not to harm others, and the quality control performed by an inclusive and reflexive science does a better job of managing that boundary than unconstrained freedom to believe whatever you want about the natural world. This is the choice that has to be made once one accepts that moral truth is the bottom turtle, that science is the next level up and that science depends on small, elite, bodies of experts. Put crudely, if you trust the science of vaccination with its idea of herd immunity, you are going to find yourself wanting to persuade those who don't trust it to take a course of action they do not like. It is not that you are harming those who don't understand it by denying them the freedom to believe what they want, it is that they are harming you and your children, and themselves and their own children, by not playing their part in maintaining herd immunity. One cannot escape the choice; to pretend it is not there is to harm the helpless – children – rather than take responsibility for their health.

How is this expanded and more reflexive science to be accomplished in practice? The solution we put forward is the idea of 'virtual diversity'. Virtual diversity rests on the techniques of ethnographic research. It requires at least one party to acquire an understanding of the other party's world. How this would work turns on the deeper understanding we now have of how sciences relate to each other. We base this understanding on the fractal model of society and the associated typology of expertises.

The ideas have been under development since the 1970s, but from around the turn of the century they have come under the heading of 'Studies of Expertise and Experience' (SEE). The basic idea, that expertise and culture are coextensive and that cultures fit a fractal model, was mentioned in the context of Figure 3.1. Developments in more recent years include the following:

a) Each group in the fractal model is associated with its own specialist spoken language. The social 'groupness' of any set of persons can, in principle, be tested in an Imitation Game, which asks a non-member to pass as a member in a verbal comparison judged by a fluent and demanding interrogator. If passing is demanding and failure is at least possible, then this is a group with its own language not just a set of individuals. One does not have to carry out Imitation Games to use the idea as a thought experiment in defining groups. Thus, gravitational waveform calculators and parents are groups; shoelace wearers and brown-haired people are 'sets' not groups.
b) Group languages are usually associated with a range of practices. Sometimes these will be narrow technical practices, sometimes just the ubiquitous social understandings arising out of widespread practices that make properly designed Turing Tests so hard for computers to pass.
c) These languages are not always inter-translatable but they can be understood through extended immersion in the spoken discourse. It is possible to understand more than one mutually untranslatable language.
d) Fluency in such a language is called 'interactional expertise'. Acquisition of interactional expertise does not confer the ability to carry out the associated practices, though it does confer enough understanding of the practices to

make sound judgements and to coordinate diverse practices in a division of labour. Ability to practise is known as 'contributory expertise', though the boundary between interactional and contributory expertise can be fuzzy, for example, when the medium of the practice is, itself, linguistic.

e) One reason for the fuzziness of the boundary between interactional and contributory expertise is that the ability to converse fluently in a domain's technical language – to have interactional expertise in respect of the domain – is to have experience of something that others do not have; it is coextensive with understanding the practical experiences pertaining to the domain even though it does not confer the ability to directly execute the practices. It does, however, enable its possessor to do things, such as make technical judgements, that others without the same level of understanding cannot do. That is why technical managers, who do possess interactional expertise but cannot carry out any of the practical procedures pertaining to a domain, can nevertheless *contribute* to it. These distinctions are important if the relationship between human biology and culture is to be understood.

The difficulty of inter-translatability creates problems of many kinds. The problem of interdisciplinarity in the sciences is one of them. This is often referred to in terms of 'trading zones', with the original discussion concerning the relationships between technicians and scientists in physics. Another example is the way chemistry and biology combined their languages and practices to form the newly merged discipline of biochemistry.[3] One way of resolving these problems without a full cultural merger of the disciplines is the 'ambassadorial model': one or two ambassadors immerse themselves in the discourse of the other group and acquire interactional expertise. They are then able to represent, if not explain (explanation requires translation), the thinking of the other group to those in their own group. Though the ambassadors may not be able to explain or execute practices, they can say what needs to be done to keep the others content, and they may also be able to coach others' learning of alien practices and languages.[4]

This kind of thinking leads into the resolution we are putting forward. The more diversity in the scientific community the better, but that is a solution to a problem of social injustice. It would only be a solution to our problem if generalized social injustices led to scientific problems – epistemic problems or problems of knowledge – but they don't, at least not in the vast majority of cases.

To make use of the example of the Maori, they are nearly always unrepresented in the profession of science whatever the topic but this doesn't lead to epistemic problems – for example, physics would not work better if there were more Maori physicists in university physics departments. It may be that it would be more justice if there were more Maori physicists in university physics departments (though even this is not clear in countries outside of those where they are well represented in the population), but no one seems to think that this is a problem for physics per se. It seems from this example, and others we will encounter in due course, that the resolution of social injustice and the resolution of epistemic weakness are close to orthogonal except in specific cases such as women: the resolution of social injustices usually doesn't lead to the resolution of epistemic problems.[5]

The reason for this is the narrowness and closed worlds of scientific specialties. Consider the case of social injustice that, for better or worse, has gained almost mythical status, the case of the post Chernobyl Cumbrian sheep farmers. According to Wynne, the farmers' experience-based expertise was excluded from the scientific debate about the treatment of grazing sheep consequent on the irradiation of the Cumbrian fells because of the farmers' low social and scientific status. But let us imagine that British society was evened out in the fashion of a socialist utopia such that sheep farmers were now a high-status group appropriately represented according to their numbers in the population in a prestigious profession such as scientific research.[6] They are still a small group of experts, not much larger than radioactivity experts and the odds against an ex-sheep farmer from the Cumbrian fells, or such a sheep-farmer's offspring being (a) a radiation scientist and (b) among the radiation scientists involved in the particular analysis of the post-Chernobyl Cumbrian fells, are vanishingly small, so there would still be almost no chance of there being any tradition of sheep-farmer experience among the scientists. And even if the sheep farmers were welcomed into the discussion in consequence of their high status, the problem of their lack of scientific socialization and, therefore, ability to converse with the scientists, would still have to be solved via interactional expertise: one or more of the scientists would still have to enter the sheep farmers' world and, like ambassadors, bring their tacit understandings of sheep ecology back into the science that was informing policy-making.

The same applies very obviously with rare ailments, particularly where they occur predominantly in old age, or where they have high fatality rates. In these cases, the chance of being treated by a doctor who has experience-based expertise in the particular ailment is almost zero, whatever the demographics of the medical profession. Nearly all the experience you are going to get applied to your medical treatment is going to come via interactional expertise: doctors learn about nearly all the conditions you are likely to suffer from by talking about them – perhaps with other doctors, often with expert patients who are not doctors – but rarely, if ever, from personal experience. Insofar as doctors are experience-based experts, it is experience in medical settings that they possess, not the experience of expert patients.

The same usually applies even where the disadvantaged group in question is large. Consider women. Gynaecology will benefit in terms of the scientific knowledge that it gives rise to as demographic dominance in the profession shifts from men to women (as seems to be happening), because it means women undergoing gynaecological procedures are much more likely to be treated by specialists with first-hand experience of the problems. Nevertheless, for many gynaecological conditions the chances that even a female doctor will have experienced the particular patient's ailment remains small. Where those ailments are rare, such as molar pregnancy or bicornuate uterus, they are vanishingly small. This is not to say that a medical profession that reflects either the wider society or its particular patient group more closely is not a worthwhile ambition. The point is simply that solutions to social injustice won't necessarily resolve scientific knowledge problems, even if they make the environment more hospitable to specialists from the disadvantaged groups.

This analysis leads into our proposed solution for most of the problems that are normally laid at the door of lack of diversity within the profession of science. The proposed solution is virtual diversity via the acquisition by scientists of interactional expertise pertaining to the diverse groups whose problems they are addressing. Ari Rip, in his discussion of the indigenous expertise of Maori such as is presented in his 2003 article, like any other social scientist, acquired (so we assume) at least a degree of interactional expertise in respect of Maori culture in order to do his social scientific work. In the same way, we are saying that scientists, where they cannot recruit diverse peoples into their research as full-blown research collaborators, and there are very few cases where they can, have a duty to acquire the interactional expertise pertaining to those whose experience they need to meld into the scientific analysis. The radiation scientists dealing with Cumbrian fall-out would be a typical scenario. This is one way we would extend the accepted norms of the scientific profession: scientists must become ethnographers on top of their other duties.

Interactional expertise and virtual diversity

We are, then, going to make the argument that the diversity needed within the scientific community can be, often is, and often *can only be*, achieved through virtual diversity based on the acquisition of the relevant interactional expertise, which means listening seriously and carefully to what others have to say.

Thus, as intimated, if, for example, you are entering a hospital for chemotherapy, the experience base of either the doctor or nurse when it comes to either the condition that is being treated, or of the experience of receiving chemotherapy treatment, will almost always consist of interactional expertise acquired from colleagues, patients or acquaintances. This is because where there are many domain-specific perspectives and standpoints (in this example, the standpoints represented by having experience of particular medical conditions and/or treatments), direct representation of each one by scientifically socialized members of each group is statistically almost impossible so virtual diversity will, of necessity, be the only way forward. This suggestion, note, resists the ever more popular claim that social understanding and representation is irretrievably tied to embodied experience (or even biology), and not language. The approach of SEE, which stresses interactional expertise, is at the other end of the spectrum and therefore represents a political intervention as well as an argument about knowledge.[7]

We are not alone in this rejection of identity politics as a rationale for insisting that only direct representation will do.[8] Even amongst philosophers of science who argue for the importance of standpoint epistemology in widening the sources of criticism that contribute to the 'objectivity' of science, there is a recognition that contributions to an expanded analysis should not be limited to those who share body types or social groups. For example, feminist philosopher Sandra Harding reflects on her own work as follows:

> Standpoint theory is not calling for phenomenologies of women's worlds, or for ethnocentric (gynocentric) accounts. Nor is it arguing that only women can generate feminist knowledge; it is not an 'identity politics' project. Men, too, can learn to start their thought from women's lives, as many have done.
>
> (Harding 1995b: 343)

Here Harding is accepting what we are calling virtual diversity.

Virtual diversity may be a new term, but it describes a traditional and very familiar methodology and a widespread type of social relationship. Virtual diversity is everywhere – the glue of society – even though this is not much discussed in academic or activist circles, where the emphasis is, nowadays, more on the difficulties or even impossibility of taking the view of the other rather than its necessity in a well-ordered society and the conditions that make it possible. Taking the view of the other is fundamental to the social sciences, such as anthropology and ethnography, and its vital role in societies is made more obvious when thought of in the context of the fractal model. Virtual diversity is what enables the various cultural domains of society to cohere into *a society* even before the failures can be criticized. Unrecognized failures of virtual diversity and weak interactional expertise do cause many well-known and heavily discussed problems, but the emphasis on these can make it seem that there is nothing to society but isolated and self-interested groups.

In terms of the fractal model, we can start at the small scale to explain the necessity of virtual diversity. The division of specialist labour depends on virtual diversity. Complex division of labour depends on the coordination, and therefore mutual understanding, of the work of others when the persons who have to understand others' work cannot do the work or, at least, have not done the work (see Figure 5.1). Since cultures at every level nearly always involve distinctive practices as well as specialist languages, and the specialist languages will make continual reference to the practices, it is convenient to refer to these sub-discourses as 'practice languages'.

Figure 5.1 illustrates the role of interactional expertise in the division of labour (the figure was originally prepared to represent gravitational wave physics, but this is not important). The figures with hammers and anvils are contributory experts with narrow practical skills, who coordinate their actions with others in the same overall specialist domain through a shared practice language (the wavy lines). The smiling figure at the bottom is meant to represent a social scientist (or, in the context of this chapter, a scientist from another field – perhaps an ambassador), who has learned to share the practice language (has interactional expertise) but can perform none of the practical skills. Moving up the fractal model, less specialist shared languages are the way higher level less specialist groups, represented by the mid-level or even higher ovals in Figure 3.1, coordinate their contributions to society as a whole.

Virtual diversity is, as already indicated, also the stuff of the social sciences. Without virtual diversity, every criminologist would have to be a criminal and there would be no such things as sociology, ethnography and anthropology where members of one social group must reach into another in order to understand it.

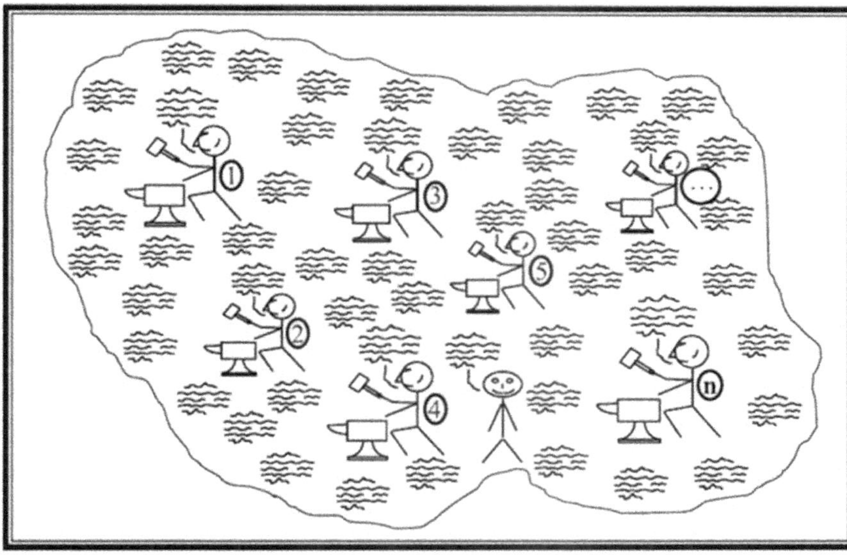

Figure 5.1 Contributory experts '1 … n' coordinate their efforts in a division of labour through a common practice language. The smiling stick-figure is an interactional expert fluent in the practice language but with no practical skills. Originally Collins (2011: 2).

When we move to the law, there would be no justice, only 'justices', each justice serving its own isolated community of understanding.

To see the problem at the most general level, Newtonian physicists wanting to understand relativity have the same problem as anthropologists wanting to understand an isolated Amazon tribe; toxicologists wanting to understand whether farmworkers are in danger when spraying powerful herbicides have the same problem as male doctors (and most female doctors) working as gynaecologists; social policy specialists trying to understand pensioner poverty have the same problem as architects trying to design prisons; and novelists trying to understand computer game addicts have the same problem as childless people trying to understand the parents of newborns. There are as many such problems as there are ways of combining any two ovals in the fractal model: if there are 'n' social groups in a society, then the number of possible combinations and the number of possible problems is n(n - 1) – just a little less than n-squared.

Moving to the arts (which is also the main topic of Walker's 2022 piece referred to in note 8 in Chapter 5), any author seeking authenticity rather than to execute a work of pure imagination needs to engage with the worlds they are describing as far as they can. Without this, individuals would be stuck with science-fiction like invention, or isolation within their own worlds of practice.

As explained, without virtual diversity, science as an institution would cease to exist because the central tenet of universality – that no scientific claim is ruled out because of the ascribed characteristics of the scientist – requires the ability to share perspectives irrespective of native culture or physical inheritance. Without virtual diversity, the clamour for social justice would destroy science. We now know, of course, that the idea of universality in science has to be thought about much more carefully given the importance of experience-based expertise but the idea that only certain kinds of people could ever do certain kinds of science would be disastrous should it become the default position: instead of science we would have a series of sciences belonging to different groups – standpoints would mean petrification in social niches rather than greater shared objectivity.[9]

Without interactional expertise, interdisciplinary work within science would be impossible without a full-scale merging of the disciplinary cultures; distinct sciences would have to merge or, in the case of outsider groups, they would have to become scientific specialists in the domain in question as well as bringing their standpoints to bear. In those cases where experience-based expertise from outside the scientific community is absorbed into the science, then the diversity is usually, and most appropriately, managed by the scientists themselves acquiring the linguistic fluency needed to create virtual diversity.[10] It is only in rare cases that the scientifically unqualified can acquire sufficient technical and cultural understanding of science to join the community and, as already intimated, it is hard to develop the trust required in scientific work if one of the cultural parties has not had a scientific socialization. These things are not impossible, but they are hard and there are many pitfalls on the way. It would be wrong to make them still harder by insisting on a full merger of cultures every time.

Scientists must choose their values

Douglas, it will be recalled from Chapter 2, is concerned that a scientific socialization is an inadequate resource for non-epistemic values that must be injected into science. To repeat her remark:

> Many of the most central values scientists hold are those they developed whilst training to be scientists, and this alone can create divergences between the

> scientific community and the general public … Thus, even a demographically diverse scientific community may still hold values that clash with the broader public's values.
>
> (Douglas 2009: 122)

But which of those values found among the broader public is to be selected for insertion into scientific practice since some of those values will assuredly be malign, either because they are something like conspiracy theories or because they pertain to a time when societal values are malign. As we will see, even Harding, a champion of diversity, agrees that we don't want Nazi values inserted into science.

There is no infallible solution to this problem, but we think that, contrary to what Douglas claims, given that someone has to choose the values that belong in science, scientists are the best bet. We should not be worried about their socialization but supportive of it because of the formative values of the institution of science with truth at its heart. Remember, this is a best bet, not an infallible recipe; it is just that there isn't a better bet. Drawing on Churchill's sentiment about democracy, leaving it to scientists to choose their values is 'the worst solution except for all the others'.

Scientists are going to have to choose among the plethora of candidates for the experience-based expertises that belong in science (in a better world, supported by social scientists, with the perspective of 'The Owls'), and there is no reason why we should not take societal values as a form of experience-based expertise – the (ubiquitous) values that the public have developed as a result of their experience of living in and developing their society. But, once more, someone must choose between the plethora of candidate values and, again, scientists are the best bet; scientists are already choosing what they want to include so as to improve the epistemic face of science so having to make choices is already part of their world. And, of course, they can always choose to draw on the expertise of other, more specialist groups, such as bioethicists, treating them as experience-based experts and applying virtual diversity once more.

The reach of interactional expertise

Language is central to culture. The approach taken here is that, in principle – though not always in practice – *fluency in any human practice language is attainable by any other human through long-enough and extensive enough immersion in, and interaction with, the spoken discourse*. The conditions for acquiring interactional expertise in other cultures in addition to personal ability, enthusiasm and energy include: (a) the group must have a rich language pertaining to the aspect of expertise that is to be acquired; (b) the hosts must be willing to allow the acquirer to immerse themselves in the discourse – to join in their social lives; (c) the acquirer must have the resources to maintain presence in that group for an extended period and must be able to return unscathed. For the extreme case, just imagine a newborn snatched from its parents in one tribe or local group and brought up in a hostile tribe.

Assuming the child was accepted into the new location, can we doubt that however different the source and host locations, and however different the baby's appearance, the child would acquire full fluency in the natural language (or practice language) of the hostile adopting community but for practical reasons could never acquire the language of the natal group.

The default assumption must be that among human language-speakers, biological characteristics are only coextensive or near-coextensive with social boundaries as a matter of historical contingency – brown skin, 'yes', brown hair, 'no'. The Imitation Game test of groupness is linguistic, however, so it does not imply that biology cannot have an influence on *contributory* expertise (e.g. only biological women can give birth).

This is a strong claim: it says that, *in principle*, among humans, so long as a practice language exists, there are no experiences or cultural capabilities the understanding of which is ruled out by biological differences between human groups or by the differences in experience between different groups. If humans are capable of talking about their experiences, as they nearly always are, all experience-based expertises are permeable to anyone given good will on all sides, enough time and the necessary resources. There may be cultural obstacles supported by biological differences but given enough determination on both sides of the divide, these can be overcome in principle; in terms of mutual understanding, biology is always subservient to culture. Thus, women could always understand men and men could always understand women if both sides want it to be the case and put in enough effort and resources, and Blacks could understand whites and whites could understand Blacks if both sides want it to be the case and put in enough effort and resources.

Properly designed Imitation Games with appropriately experienced populations ought to reveal these possibilities, which would support the contemporary view of the permeability of the boundary between men and women. To repeat, however, this does not mean that biological men can give birth to babies: that's clearly a matter of contributory not interactional expertise.[11]

The implications of this argument for interactional expertise are nicely illustrated by the controversy around Rachel Dolezal, the white woman who claimed to be Black and who, in 2014–15, came to be chapter president for the *National Association for the Advancement of Colored People* (NAACP) in Spokane Washington. Dolezal deceived her Black followers and everyone else by claiming a heritage that was not hers, and this is inexcusable, but our question is whether there is any *epistemic* obstacle in the way of a white person with sufficient interactional expertise representing a black group effectively and authentically. The answer, according to the idea of virtual diversity, is that there is not, providing that the person was prepared to put in the work needed to acquire the necessary interactional expertise and could find enough persons sympathetic to the aim to allow him or her to be immersed in their discourse for long enough. Of course, an acknowledged white person in such a role might not be able to offer effective political leadership: here, however, we are dealing with epistemic rights and wrongs, not the role of identity in politics.

A similar point can be made using Reni Eddo-Lodge's *Why I'm No Longer Talking to White People about Race* (Eddo-Lodge 2020), which was raised by a critical referee of an earlier draft of this argument. Eddo-Lodge describes the incomprehension she experiences when talking to white people about the prejudice Black persons encounter in their lives. And yet Eddo-Lodge herself accepts the possibility of virtual diversity. She says: 'I'm no longer engaging with white people on the topic of race. *Not all white people*, just the vast majority who refuse to accept the existence of structural racism and its symptoms' (Eddo-Lodge 2017, emphasis added).

Later in the same, *Guardian*, piece she discusses the murder of Black teenager Stephen Lawrence and its flawed investigation by the police, but she commends the (White) Sir William Macpherson, the high court judge who chaired the subsequent judicial inquiry and found that the investigation 'was marred by a combination of professional incompetence, institutional racism and a failure of leadership by senior officers'. This institutional racism, the report explained, is:

> the collective failure of an organisation to provide an appropriate and professional service to people because of their colour, culture, or ethnic origin. It can be seen or detected in processes, attitudes and behaviour which amount

> to discrimination through unwitting prejudice, ignorance, thoughtlessness and racist stereotyping which disadvantage minority ethnic people.
>
> (Macpherson [1999], quoted in Eddo-Lodge 2017)

Macpherson is here exemplifying virtual diversity.

This understanding of the Dolezal case, the Eddo-Lodge claim and the question of embodied vs interactional expertise matter because the epistemic analysis bears on questions of representation and understanding that are central to many professional advocacy roles, including legal representation, social work, mental-health and many other welfare-related settings.

Dangers of virtual diversity

The most obvious peril of virtual diversity is that it will be done badly but still used as a warrant for exploitation. The very concept invites a 'virtual diversity lite' interpretation in which some superficial conversation with unfamiliar groups or groups with unfamiliar expertises makes the investigator feel they have acquired enough interactional expertise to speak for them.[12] If handled insensitively, especially in the case of indigenous cultures, there is a real risk it will bring out fears about colonialism and domination: 'here is the outsider, purporting to understand our society and telling us how we should live our lives'. But the solution is not to abandon virtual diversity but for the scientific community to work within today's understanding of these problems and handle it assiduously and sensitively. If we want scientific medicine and the other benefits of science, political as well as substantive, we have to be vigilant and ensure that virtual diversity is always attempted with integrity and never expediency. There is no principle that prevents inclusion and representation being achieved along with epistemic authority.

The other danger is that the idea of virtual diversity falls into the same trap as the argument for democratizing science and becomes just another way of including an indefinite range of cultures. To avoid this, we need not only to remember that interactional expertise is a hard thing to acquire but also that there are a limited number of experience-based expertises that merit acquisition if the intention is not just to explore other cultures but to meld them into a specific field of scientific activity. This, once more, is what we have referred to as the problem of extension.

Here we echo feminist philosophers of science such as Harding and Longino, both of whom champion the inclusion of wider groups into the scientific process in order to strengthen it, while realizing that it could also destroy it. Harding argues that 'stronger objectivity' can be created by the inclusion of more standpoints reflecting diverse opinions in the scientific debate, while Longino agrees that 'the greater the number of different points of view the more likely it is that scientific practice will be objective … [and] … reliable' (Longino 1990: 80). Crucially, however, both recognize that expanding the number of opinions to be absorbed into science, however it is done, must be done with a reason and a limit in mind. Thus Harding:

> Obviously not every starting point for thought that lies outside a dominant conceptual framework is likely to enlarge our understandings. We can agree with the defenders of weak objectivity that at least some of the interests and values they think should be excluded from directing knowledge projects do indeed retard the growth of knowledge – 'Think of Nazi science!'
>
> (Harding 1995a: 25)

As far as we can see, Harding does not pursue any methods for limiting and choosing standpoints, but Longino (2002: 129–33) does go further, attempting to solve the problem by defining four criteria that can be used to assess the 'epistemic effectiveness' of potential contributions and hence prevent what she calls the potential 'cacophony' of unchecked debate. Of these, the two which translate most directly into collective actions by the scientific community and wider society are:

> *Venues*. There must be publicly recognized forums for the criticism of evidence, of methods, and of assumptions and reasoning [and] … *Public Standards*. There must be publicly recognized standards by reference to which theories, hypotheses and observational practices are evaluated.[13]

What might this mean in practice? The first step in building a boundary around what is to be included is to handle fringe sciences. A recent case study has found that the antisemitism that persuaded the Nazis to dismiss 'Jewish science' is still being deployed to bolster the rejection of relativity in the contemporary fringe of physics.[14] The dilemma of what to reject extends to the entire fringe whether racist or not. For example, the same case study looks at the developing problems caused for the preprint server, arXiv, originally set up with intention of being open to anyone but which soon encountered the need to demarcate far-outlying views to avoid being overwhelmed by alternative viewpoints. (A space for many of these views is nowadays available in the 'alternative' preprint server 'viXra'.)

Even with the fringe scientists excluded there remains an enormous range of social groups who might claim to have a stake in the debate or some experience-based expertise that is directly relevant to it. The problem of extension is the failure of 'the democratizers' to recognize this as a problem at all.[15] In contrast, the approach taken here puts some practical detail on the sentiments expressed by Harding and Longino by making scientists (preferably including some of the right kind of social scientists) primarily responsible for recognizing and taking into account the *scientifically valuable* expertises of those who are not scientifically socialized. This does not exclude outside activists from campaigning to be included themselves or for particular issues to be prioritized – such activism is an unquestionable right in a democratic society – but it is the scientists who must demarcate the scientific sense from the lobbying and interests. To ensure that this does not simply reproduce the status quo ante we have added a new duty requiring the need to discover experience-based expertise, in the form of virtual diversity, to the norms and procedural safeguards that characterize science. Virtual diversity involves, among other things, discovering the cultural truth about initially alien groups, and, as intimated, it has to be accomplished with the same assiduousness as the pursuit of truth in the rest of science. The acquisition of virtual diversity, we are arguing, has to be treated with as much reverence and perseverance as the other norms of science even though the 'correspondence truth' involved here – understanding and telling the truth about other cultures – is a matter of social, not natural, science. In this way, the moral dilemma of who to trust is solved by giving a special privilege to an institution that has the formative intention to discover the truth, and which has learned that in order to discover truth as a community, one must tell the truth as an individual.

Illustrative case studies

We now illustrate the analysis with a series of examples, most of them already well-known within social studies of science. Each will be analysed in terms of the role of the expertise that is contributed by the cultures that come into contact, and each followed by a checklist of conclusions. Anticipating, we will find that full mergers between culturally diverse groups are difficult and rare unless both cultures have already been scientifically socialized, in which instance it is essentially a matter of interdisciplinarity; we will find that, occasionally, scientifically unsocialized persons can acquire scientific expertise but usually the onus is on scientists to acquire interactional expertise in respect of experience-based experts; we will find that ameliorating social injustices often has no positive effect on the science, one of the reasons for this, as has been explained already, is that the scientific expertise is found in narrow specialties and the experience-based expertise needed to improve it is, in most cases, a narrow specialty, and finding them both in the same place is a vanishingly small possibility; we will find that too much passion for social justice in science can have scientific consequences which lead to harm; where the prime aim is to alleviate social justice it is sure to have positive scientific impact only under special circumstances; we will conclude that such special cases should *not* be treated as paradigmatic for all science; we will find political activism can lead to positive epistemic effects if backed up by scientific understanding but to negative epistemic effects and subsequent harm where political power overrules scientific expertise.

The ordering of the numbered headings is such that the first one is a case where there is a null relationship between social injustice and epistemic consequence; headings 2, 3 and 4 exhibit a positive epistemic effect from remedying social injustice with heading 5 describing an ambivalent case; cases 6 and 7 exhibit a strong negative epistemic and policy effect from over-zealous or incorrect interpretation of injustice, both being, among other things, exemplary instances of the problem of extension. Taken as a whole, the examples illustrate the conclusions drawn from the typology of science set out in Chapter 4, namely that there is no one-size-fits-all solution to problems of widening participation in science – each case needs to be considered on its own merits.

1 Rosalind Franklin

Rosalind Franklin's work using X-Ray crystallography was foundational to the discovery of the double helix, but according to Crick and Watson, they stole it and used it to their own advantage. Many believe that Franklin should have shared the credit, and, if she had lived, the Nobel Prize, for the discovery. A recent interpretation suggests that the standard story promulgated by Crick and Watson is untrue and that Franklin cooperated fully with them, but even if that is the case, the standard account works as a thought experiment, and we will proceed as though it was the case.[16]

Franklin, on the standard account, suffered a social injustice that is indicative of the injustices long inflicted on women in science. But our question is whether this injustice undermined the quality or value of the *science* produced in the discovery of the double helix. The answer seems to be 'no'. While Crick and Watson's willingness to proceed without acknowledging Franklin's contribution may have mirrored the widespread prejudice and discrimination in respect of women in science, the quality of Franklin's scientific work was clearly recognized by Crick and Watson – that is the very reason they used her results (whether stolen or not) and, as far as we know, no-one has since claimed that Crick and Watson's claim about the double helix was flawed because a woman did or did not contribute to it.[17] Something similar might be said of almost the entire history of physics.

The main lessons from this example can be summarized as follows:

Groups involved:	Molecular biology and crystallography within science, men and women from wider society
Injustice and knowledge:	Social injustice did no epistemic damage
Politics:	No obvious effect from or on wider society at the time
Who was affected:	Established scientists
Crucial actors:	A few established scientists
Conclusion:	Social injustice is not necessarily epistemically damaging

2 Early primatology and reproductive science

Above we have explained why demographic adjustments to the population of scientists rarely solve scientific problems, namely that expert specialist groups in the case of both scientists and experience-based experts are nearly always small and unlikely to overlap. Nevertheless, there are exceptions: sciences in which the cultural position of the demographic group as a whole does have an epistemic bearing. Primatology and reproductive science are an exception to the rule put forward earlier that: '*one is unlikely to encounter a specialist in the "condition" being investigated by adjusting the entry conditions to the profession*'. These sciences are an exception because in these cases, *the 'specialist' expertise is simultaneously a ubiquitous expertise.* That means, for example, that the social injustice experienced by women within science and in the wider society can have a direct effect on the science and that this effect can be ameliorated by increasing the representation of women within those sciences.[18]

For example, early primatology studies were conducted by men and tended to focus on stereotypical male behaviours such as aggression in species where males tend to be the dominant sex.[19] The behaviour of female primates in these settings, or of species in which females played a bigger role (i.e. matriarchal not patriarchal groups), was simply not studied and behaviours observed in the patriarchal species were generalized as typical, natural and normal traits. As female researchers began to enter the field, they challenged and changed its research practices by studying different species and by paying more attention to the role of the females within the group. The result was that a much wider range of social structures were identified within primate species and a much wider range of behaviours documented. This, in turn, undermined the claims made by the male-dominated researchers that the behaviours they had recorded were natural and normal, an argument that was often put in an evolutionary context to justify and normalize similar traits in humans.

A similar addition of complexity and nuance occurred as female researchers began to enter the domain of reproductive science. Here the influence of male dominance was reflected in the gendered language used to describe the different characteristics of the egg and sperm (the classic account is Martin 1991). Thus, the female egg was seen as being essentially passive and doing little more than waiting for the sperm. In contrast, the sperm was highly active and agentic, both in the way it was seen to compete to reach the egg and also in the way that it was seen as penetrating the membrane in order to fertilize the egg. Over time, perhaps as female researchers entered the discipline or perhaps due to increased awareness of gendered language, the characteristics assigned to egg and sperm began to change, with the egg being seen as a more complex entity that played a more active role in selecting and enabling the sperm to fulfil its function. Nevertheless, gendered language continues to appear in medical and biology textbooks (Campo-Engelstein and Johnson 2014), demonstrating how difficult it can be to change deep-seated cultural assumptions.

We have treated these two cases together because they illustrate the same point which, to repeat, stands in contrast to the majority of the examples discussed and the majority of cases which could be found in science. The reason is that in these cases the specialist expertise that had been lacking and led to negative epistemic consequences was, at the same time, a ubiquitous expertise within the wider society as the excluded group is so large. As such, including that group had a good chance of remedying the problem. Just including the view of the world from a female perspective, as that perspective is most generally found, or at least represented, in Western cultures, rectifies the narrow view caused by the pre-existing social injustice.

But these are not paradigmatic cases of distorted science but very special cases of unusually gendered science. It is probably not a coincidence that these are also the sciences where much of the impetus for the early work in feminist and standpoint philosophies of science emerged. To repeat, the very influence of changing the gender balance of those working in primatology and reproductive science is an exception that proves the rule because it is a rare case of the coincidence of ubiquitous and specialist expertise related to the actual topic of the research.

The key features of these examples can be summarized as follows:

Groups involved:	Men and women within specific scientific disciplines, men and women from wider society
Injustice and knowledge:	Injustice was causing poor science until gender balance rectified it
Politics:	N/A except perhaps for internal politics of science
Who was affected:	Primatologists/reproductive scientists
Crucial actors:	All scientists in these fields
Conclusion:	Injustice and epistemic damage can be resolved by changing gender balance in the science, but only if the relevant specialist expertise is more accurately classed as a ubiquitous expertise. These cases are untypical, rather than exemplary, and should not dominate thinking about the epistemic consequences of social injustice.

3 San Francisco AIDS activists

The AIDS activists' successful campaign against double-blind tests has been documented by Steve Epstein (1996). Though Epstein claims the crucial intervention was a matter of political activism, the activists acquired enough understanding of the science to be congratulated by the scientists they were initially opposing, and we believe it was this high level of enculturation that was crucial to their success. It may well have been that the salience of the political activism caused scientists to look at the claims but without the activists acquiring enough technical and methodological expertise in the science to propose viable new protocols in place of the double-blind tests to which they were objecting, we do not think they would have been taken seriously or succeeded in modifying established practice. This, then, according to our analysis, is a case of full cultural merger between scientists and unqualified outsiders, the latter acquiring scientific socialization.

In contrast to the case of Rosalind Franklin, the expertise of the treatment activists was not immediately recognized by the mainstream medical community. Nevertheless, over time and through a combination of learning, engagement and political activism, the treatment activists were able to build alliances with those members of the medical research community who were also concerned about the design and conduct of clinical trials and, eventually, to have their concerns recognized as legitimate. The design and conduct of clinical trials were changed as a result and members of the activist community began to play a more official role within the medical research establishment.

The most important elements of this example can be summarized as follows:

Groups involved:	Medical scientists, well-educated AIDS sufferers
Injustice and knowledge:	Injustice could have had a negative outcome but didn't
Politics:	Initial injustice was rectified by political activism, but full cultural merger was also established by activists
Who was affected:	Expert patients
Crucial actors:	The expert patients
Conclusion:	Political action can be influential creating the conditions needed to recognize an injustice but a full cultural merger between initially scientifically unsocialized citizens and scientists is rarely achievable.

4 Ethnobiology

In the artisanal fishing village of Siribinha in Northeastern Brazil a multidisciplinary academic team carried out extensive enculturation and social science fieldwork looking at fishers' expertise and the existing overlaps with scientific knowledge.[20] The researchers identified similarities and differences in fishers' classification of species in respect of scientific taxonomies. They revealed, for example, that local knowledge about fish spawning periods meant that seasonal policy restrictions on fishing were being implemented at the wrong time for that location. Changing this impacted fisher's lives positively and also improved conservation efforts. Similar results have also been reported in other species and locations.

In terms of SEE, the ambassadorial work of shedding light on the traditionally neglected knowledge of fishers created the possibility of improving science through virtual diversity, bringing tangible benefits to the fishers but not requiring them to become scientists. Carrying out such work is clearly complicated and fraught with ethical complexity. The ontologies and epistemologies of indigenous and mainstream science both come with wider sets of values that may be difficult to reconcile, thus giving rise to concerns about exploitation and extraction in which indigenous knowledge is simply mined for whatever the dominant science can use. Nevertheless, this case shows that given sufficient care and reflexive awareness, the norm of virtual diversity can bear fruit.

The key lessons for future practice that we take from this example can be summarized as follows:

Groups involved:	Established scientists (natural, social, ethno and philosophy) and local experience-based experts
Injustice and knowledge:	Injustices rectified through virtual expertise with positive effects
Politics:	Concerns about sustainable development put focus on indigenous communities and their role in managing and preserving ecosystems
Who was affected:	Scientists, local fishers
Crucial actors:	Ethnobiologists and philosophers acting as intermediaries and ambassadors creating virtual diversity

Conclusion:	Virtual diversity can be created through careful and proactive efforts that avoid 'extractive' models of knowledge acquisition.

5 Cumbrian sheep farmers

It seems clear that the Cumbrian sheep farmers possessed experience-based expertise that would have made a valuable contribution to the scientific analysis as initially conducted by scientists from the UK Ministry of Agriculture, Fisheries and Food (MAFF). Unfortunately, it was ignored but, even if it had been recognized, using that expertise would have required a crossing of cultures. The farmers were not socialized into the cultural aspirations of science and had no fluency in the science's specialist 'practice language'. Even if the scientists who were responsible for providing advice on how to handle the sheep on radioactive pastures had discovered that the sheep farmers had something to offer, it would still have needed melding into their scientific culture and in this case full cultural merger would be an enormous undertaking for the farmers. In cases discussed under the second heading above, cultural merger was made much easier because, at the outset, both groups shared the language and culture of science as a whole. The sheep farmers did not possess that fluency nor is it likely that they would have the time and inclination to acquire it by undergoing an apprenticeship in science. If this cultural divide was to be crossed, it would depend on the MAFF scientists immersing themselves in the language of the sheep farmers and absorbing their understandings into the science. It would not require them to become sheep farmers, only to develop a level of interactional expertise. This level might be quite shallow since the mistakes that the scientists were making in terms of sheep husbandry might not rest on much more than *information* about how sheep behave and the economics of sheep husbandry.[21]

The main features of this case study can be summarized as follows:

Groups involved:	Established scientists, experience-based experts (sheep farmers)
Injustice and knowledge:	Injustice (in the form of class prejudice) did lead to negative scientific outcome, but it looks as though it could have been rectified easily with virtual diversity.

Politics:	Generalized distrust of government scientists by local community as a result of past experience with nuclear industry that increased as post-Chernobyl restrictions changed without warning or explanation
Who was affected:	Both groups, though sheep farmers suffered most as their livelihoods were directly affected by policy changes
Crucial actors:	MAFF scientists failing to acquire virtual diversity; sheep farmers as relevant experience-based expertise
Conclusion:	Checking for relevant experience-based expertise in order to create the virtual diversity needed to ensure epistemic standards should be the norm.

6 MMR

Where an existing consensus is challenged, this needs to be on the basis of evidence and expertise and according to the norms and values of science. In the previous examples, these characteristics were met either by the activists themselves or by the ethnobiologists and philosophers acting as a bridge between the two communities. In the case of the measles, mumps and rubella (MMR) vaccine, these conditions were not met. Instead, we see a potential insider who fails to uphold the expected standards and a range of outsiders who lack the in-depth knowledge possessed by the treatment activists, sheep farmers and fishers.

In the late 1990s a medical doctor called Andrew Wakefield claimed in a press conference that the MMR vaccine could cause autism. The impact of this claim was aided by *The Lancet*'s publication of his article (Wakefield et al. 1998), which was the occasion for the press conference and which was not retracted for another twelve years, and by the mass media's preference for stories that mistakenly 'balanced' scientific evidence with parents' claims about the cause of their children's autism. The net result was a widespread anxiety amongst parents and a significant drop in the proportion of parents taking their children for the MMR vaccine, which has led to repeated measles epidemics in many locations across the UK and which continues to this day. But though some, or even many, of the parents who supported Wakefield witnessed the agonizing onset of autism in their child shortly after an MMR vaccination, there was no evidence of a causal link even at the time. A reading of the original *Lancet* article reveals that it never makes this claim, not least because of the totally inadequate statistics, and all the available epidemiological evidence was against it. Later it would turn out that Wakefield had a financial interest in his claim and that his recruitment of cases was unethical, at which point the paper was withdrawn but most of the damage had already been done.

Sadly, a number of prominent figures from the STS field endorsed the parents' view, which was very popular at the time in such circles and helped to give rise to another example of something close to a ubiquitous science. Still more sadly, to repeat a point made earlier, as far as we know, now that the reality of the dangers of the revolt has revealed themselves in concrete form with huge increases in outbreaks of measles, none of the social science-based protagonists have reconsidered their one-time involvement and reflected upon their mistake; this damages the credibility of our discipline as a determined seeker after truth.[22] Worse still, the legitimation of a revolt against one vaccine is likely to be seen as supporting other vaccine revolts. This is why it is so important to disentangle sentiment for those outside the scientific elite – often widely admired among social scientists – with actual epistemic damage. The epistemic damage in this case was a consequence of the uncritical championing of the views of actors who were marginal to the science but who were also the most important stakeholders.

The crucial lessons from this example can be summarized as follows:

Groups involved:	Medical scientists, citizenry/social scientists/commentators, who all treated themselves as having the same epistemic rights as accredited scientists (problem of extension)
Injustice and knowledge:	Imagined injustice led to negative epistemic outcome and greater injustice
Politics:	Distrust of politicians and science advice based on earlier controversies that was reinforced by media representations that typically presented the issue as a 'balanced' debate
Who was affected:	The defenceless, such as children and subsequent generations
Crucial actors:	A small group of citizens (plus social scientists, celebrities, journalists)

Conclusion:	Imagined injustice can be dangerous as small groups can exert a disproportionate effect. This means elites, including the media, governments and scientists, have important roles to play in communicating what is and is not known and holding each other to account.

7 Anti-retroviral drugs in South Africa

A similar case of unhelpful and uninformed 'participation' in science having a disproportionate and devastating effect is the decision by South African President Thabo Mbeki *not* to distribute anti-retroviral drugs to the population at the time of a growing AIDS epidemic. Mbeki claimed that scientific research showed that the drugs were unsafe and inefficacious.[23] In fact, while a debate about the drugs' efficacy and safety could be found on the internet, the scientists putting the anti-case were no longer able to publish their views in regular journals: Mbeki's 'scientific' support came from the scientific fringe.

In a backhanded way, the Mbeki case supports Epstein's view of the importance of political power in the case of science. Mbeki and his colleagues were running South Africa and, unsurprisingly, they got their way, with, it is now understood, seriously damaging effects on, for example, babies born to mothers not given the drugs: the babies were born with AIDS whereas the drugs could have avoided or reduced the incidence of this outcome. Here there was no full cultural merger between the established scientists and outsiders (including the fringe, some of whom were at one time well-established members of the mainstream but no longer). Nor was there any virtual diversity with the political establishment either failing to acquire, or failing to use, interactional expertise in respect of the established science of anti-retrovirals, instead presenting the fringe as the establishment. Mbeki acted as though, like Trump and the MMR revolt supporters, his scientific understanding was superior to that of the scientific establishment.

The lessons we take from this example can be summarized as follows:

Groups involved:	Medical scientists, political leaders
Injustice and knowledge:	Failure to correctly identify mainstream scientific consensus led to poor political decision-making and misleading justification of policy choice

Politics:	Justified concerns about public health within South Africa given the number of infections, but also wider issues of national reputation and costs of treatment, coupled with a president who rejected mainstream medical science on HIV/AIDS
Who was affected:	Pregnant women and their unborn children who were put at greater risk of developing HIV/AIDS
Crucial actors:	Political leaders who failed to recognize relevant expertise; opposition from mainstream science and civil society that eventually forced policy to be changed
Conclusion:	Those outside the scientific community may struggle to recognize current consensus, with the result that too much weight is given to marginal or out-of-date ideas. Expertise is needed to recognize what current consensus is and what counts as reasonable questioning of this.

The wider perspective?

It could be said that this kind of case-study-based analysis misses the larger point – the science would have turned out differently had a 'stronger objectivity' been realized through the inclusion of more diverse perspectives at the outset. That, say, Rosalind Franklin's exclusion from the story of the double helix was not just a matter of null epistemic consequences but of the androcentric culture of science being further reproduced and legitimated. Could it be that if there were more women in science even physics would look different or, at least, dominate the 'softer' sciences less?

These concerns are important but are close to orthogonal to the argument made here. As noted at the beginning, and as evidenced in the examples, moves to address social injustice and attempts to improve epistemic quality rarely overlap and do not necessarily pull in the same direction. We take this to mean that addressing social injustice by, for example, improving the diversity of the scientific community must be seen and justified as an end in its own right and not as a means to some other end, such as improving epistemic outcomes. There will be some cases where the two goals align – hence the 'close to' orthogonal – but these are the exception, not the rule, so improving demographic diversity and representation needs its own rationale.

If demographic representation were to be improved, it would have an effect on the progress and priorities of science but the argument for virtual diversity means it would not be the best that science could do. The reason is that, as the examples have shown, the diversity that is needed within most science – be it pure, applied or policy-driven – is too varied and too context-specific to be found within the specific group of scientists who make up the relevant core set. In these vast majority of cases the diversity needed will have to be found from outside the scientific community and be incorporated into its members. The role of experience-based experts is only rarely to become scientists. Acquiring the virtual diversity relevant to each particular case would strengthen the objectivity of science by including more relevant perspectives, resolving specific cases of social justice rather than being driven by the goal of resolving social justice in the larger community.

Summary

We have brought out another feature of a trustworthy science, namely that when operating across cultures, scientists should put the same assiduousness and integrity into achieving virtual diversity as they put into other aspects of their work. Harding and Longino put the matter in terms of stronger objectivity arising out of multiple standpoints, but we argue that this cannot work. We stress the need to bring in relevant experience-based expertises. The new contribution to the debate is the argument that this diversity should, in most cases, be achieved not through bringing outsiders into science but by bringing in their expertises via virtual diversity managed by the scientific community. No scientific community could or should be truly representative of all human groups short of a merger of all human societies, so, if we want to continue to aspire to maintain a universal science rather than proclaim the rights of all peoples to have their own way of seeing the world, we need to decide which kinds of diversity matter and how to include them and exclude the others. Problems of cultural domination and distortion are not solved simply by making the body doing the research more demographically representative, or even by bringing in topic-related diverse groups however politically praiseworthy this may seem.

The problem of communication and trust still has to be solved and there is no merit in replacing a political problem with a scientific problem. Diversity must be handled with attention to detail for each specific topic and for each specific case, recognizing the different problems as they play out upstream, downstream and at the research front in a defined range of types of science.

Our principal solution is virtual diversity directed and executed by those committed to truth and belonging to the institution of science. This is also a proposed and partial solution to the most pressing political problem in the West since the Second World War and it grows out of the close examination of the social nature of scientific knowledge, starting with an understanding of specialist division of labour. Gender equality is always a social good, so we would want it in all of science, but it is only an epistemic good where gender is relevant (provides a specialist expertise) in respect of what is being researched. The same principle would also apply to other topics and standpoints. Virtual diversity solves this problem by enabling specialist expertises to be included even when social justice has not been achieved and when the most difficult conditions for full mergers into the scientific community have not been met.

6

Economics and ubiquitous economics: How not to be a fulcrum science

Introduction

Our overall aim is to show how science as an institution can be a foundation for veritocracy. We have argued that science is an institution in which truth is central to the formative intentions and that this moral imperative has the potential to feed back into society as a whole. But this applies only to sciences that share the formative intention – fulcrum sciences – and by the time we had worked through to Chapter 4 we had seen lots of exceptions.

One thing we are doing here is revisiting the problem of demarcation – how to demarcate science from non-science. As has been seen, our approach is *not* to think in terms of a set of necessary and sufficient criteria for being a science but to accept the kind of fuzziness introduced by Wittgenstein when he discussed the meaning of 'game', and to persevere nevertheless. As we put it in our (2017b), *Why Democracies Need Science*:

> Luckily, science only seems hard to define. It seems hard because a mistake has been made about what it is to define something like science. Wittgenstein pointed out that, even though we regularly use the word and the idea of 'game' without being puzzled, we cannot define a 'game'. (p31)[1]

Our approach may not allow us to define every action we encounter as either inside or outside the category of science, but it does allow us to exclude a lot of actions. Something is not science if the intention is not dominated by the aspiration to discover correspondence truth and act with moral truth – that is a necessary condition but not sufficient. For the sufficient conditions one must refer to the rightmost column in Table 1.2 and much more in the way of the 'game-like' qualities of science as an institution. Not all of these aspirations are necessary conditions, and science can proceed while a number of these aspirations are not met, but a family-resemblance related handful of them looked at on a case-by-case basis are sufficient.

In this chapter we are going to look at economics as an example and try to apply our demarcation procedure to it. Contrast this with, philosopher, Nancy Cartwright talking about the demarcation problem in economics:

> The long and the short of it is we have not made much progress on saying what science is after 60 years of serious work in the philosophy of science ... This past autumn [this probably refers to autumn 2014 or 2015], the Philosophy of Science Association, at their big international biannual meeting, after many, many years of the issue lying dormant, had a session on 'What is the Scientific Method?' Nothing came of it. So it is a little hard to answer the question 'Is economics a science?' My answer is that I do not have a clue.
>
> (Cartwright and Davis 2016: 44–5)

In contrast, here, we will provide some clues.

One important kind of exclusion in the case of economics is ubiquitous science, where assumptions are still being debated in the core, but they have a public face in which these debates are suppressed helping to give the public the false impression that their ubiquitous expertise encompasses the technicalities of the science and engages with decision-making at the research front. Another exception, this time within the academic domain, arises out of the reification of models, where the manipulation of models takes over from the assiduous search for correspondence truth. Both kinds of activities can be found within economics or associated with economics. We concentrate on the United States and the UK, though the influence of the economics emerging from these countries is wide.[2]

'Economics' covers a range of activities. We have already mentioned microeconomics and macroeconomics and the characteristics that lead them to be located in different rows of Table 4.1 and Table 4.2, while ubiquitous economics, which will feature large in this chapter, is located in the interstitial category of those tables along with reification of models which takes place inside universities as well as outside. 'Academic economics' is found in universities and includes micro- and macroeconomics but also 'mainstream' and 'alternative economics' which fight it out within academic departments. There is also 'free-market economics' which is found in universities and influences the mainstream more than the alternative stream; all these kinds of economics will be encountered in what follows. We will argue, among other things, that economics' enormous success in using mathematics to build theories based on a stripped-down model of human action has been counter-productive in terms of the assiduous pursuit of correspondence truth. The very success of the theories drags even the economics located in universities in the direction of theory and away from the very much more complicated analysis of action in real societies. Even 'behavioural economics', which tries to rectify some of this tendency by investigating real human behaviour under experimental conditions, works with small numbers of people engaged in 'toy' tasks in isolated laboratories. We'll contrast economic models turning on the idea of freedom of choice, with the complexity of the causal influences on action found in real social life. Still more marked is the simplification that occurs when economics escapes from the academy and the way this has led a typical human science found in universities, with all its complexities and disagreements, at least mentioned, if soon forgotten, to become a formative part of public and political life. We do not mean to say that an economics that could be taught in universities, and may, occasionally, actually be taught in universities, and practised in real-life situations, with all the fundamental disagreements revealed along with the many exceptions to the mathematically established conclusions, would not be a fine science and a fine academic subject, especially as it is promulgated in those places where the hegemony of the market-obsessed neoliberalism has been replaced by less ideological concerns.[3] Though consensus is the aim of science, in social sciences it is often disagreement that is the touchstone of integrity. Where there is too much agreement in sciences characterized by disagreement over fundamentals, it can be an indicator of the pursuit of something other than truth.

Economics in the United States and UK, which we'll call Western economics is an excellent subject for a case study relating to our topic, not only because *large parts of it* (not all of it), are an exception to our definition of science as the assiduous pursuit of correspondence truth, but because it illustrates both the left-hand and right-hand panels of Figure 3.2. Western academic economics has been influenced by societal culture (the left-hand panel). In particular, as we will see nearer to the end of the chapter, Western academic economics has been affected by that part of the culture that has been influenced by big business, while ubiquitous economics has influenced society and its politics (the right-hand panel). There is also a feedback loop – society influencing economics, economics influencing politics and national culture, and that national culture creating a still more favourable cultural environment for the kind of ubiquitous economics that filters up again into society. One mechanism is what the young are persuaded to take for granted in their society and how this will influence the way they interpret any conflict they encounter in their university experience ; another is the way governments and sponsored think tanks will favour economists who share their view of economics in consultancy jobs and so forth.

We will argue that, as a result, these parts of economics have been one of the major contributors to the erosion of truth in Western societies that we want to reverse; Western economics, through its opposition to the central control of economies, its advocacy for markets over professions and its insistence that government intervention inevitably leads to left-wing totalitarianism, has contributed to the rise of populism and the demise of veritocratic politics, pushing society towards *right-wing* totalitarianism.

Ironically, looking at the role of economics since the Second World War, we can see that it *is* possible for a scientific field to shape a national culture in profound ways. This provides backhanded support for our major thesis that the right kind of sciences could change national cultures in the right kind of way. Economics as it is currently practised is not a suitable science to be a foundation for veritocracy, but the way it has affected society does indicate the kind of effort that would be required for a science to re-orientate society's values. So, if there is one science that can be taken as a case study to illustrate many of our arguments, including the special role of ubiquitous sciences, economics is the obvious one.[4]

Democratizers, scientific value analysts and ubiquitous economics

Again, the way this feedback loop with society plays itself out in the case of economics ought to provide an excellent case study for the democratizers: the way economics works is surely what the democratizers want from a science! They want a science which invites the participation of the public and feeds that participation back into the scientific institution with governments talking science to citizens, scientist-to-scientist, in the style of equals: neoliberal economics is the perfect example of this. Thus, politicians and citizens together consider whether taxing the rich will make the country poorer; politicians and citizens debate whether free markets will make the country richer; politicians and citizens debate whether private enterprise is more efficient than government provision for services such as transport, water supply and waste disposal. Governments seem to have accepted that 'There Is No Alternative' (with its own acronym TINA), to free markets and 'neoliberalism', and citizens, armed with their understanding of ubiquitous economics, have largely agreed.

Ironically, since technocracy is anathema to democratizers, this is a clear case of technocracy – a 'scientific' theory leading the political choices of the government by the nose. The circle is squared because the citizenry participates in the science: it is technocracy via the will of the people![5] Here it is the left-hand panel that is most illustrative: public discourse and public values in the form of neoliberal economics, influencing the science.

Representatives of these positions might want to react differently to the implications of what has just been written, but if they want to reject them, they, like us, will have to invent criteria that distinguish between societal values that should inform science and those that should not. The solution we offer has been set out in Chapter 2: the decision about which values should influence science should be relegated to technically informed and trusting groups sincerely aspiring to locate correspondence truth – in other words, scientists (though, as set out in Chapter 5, they may choose to incorporate other sources of expertise via virtual diversity).

Academic economics, liberalism and libertarianism

We now turn to the science of economics that promulgators of ubiquitous economics draw on and simplify so as to shape the wider culture. In academic economics the 'mainstream' is buttressed by a well-established hierarchy of academic journals, while 'alternative' economics (otherwise known as 'heterodox' or 'pluralist' economics), continually questions the assumptions upon which the mainstream is based. Such disagreements are normal and even desirable in a healthy social science, with different groups of actors each believing they have the key to correspondence truth. In this case, however, heterodox economics finds it especially hard to change mainstream thinking and its supporting institutions.[6] It appears, as will be documented below, that the balance of disagreement between the competing positions in academic economics has been influenced by outside interventions from interested parties.

There is nothing new about the 'internal' critiques of economics we discuss below, and the historical sketches of events are also derived from others:[7] the new contributions are the effect on the balance of the argument in academic economics, the idea of ubiquitous economics and aspects of the relations of these things to wider societies. We draw on existing critiques to argue that much of the mainstream of economics is structurally unsuitable to be a foundation for democracy because it is founded on a misplaced model of science, and also because economics, often against the will of the minority of academic economists who are classed as 'alternative', evolved as a contributor to public life and societal values in ways that have taken it away from the formative aspiration of the search for correspondence truth. By claiming that it is the pre-eminent value-free social science, or having this status imposed upon it because of its mathematical carapace, it has been transformed into a powerful generator of values for the wider society. The critiques that follow start with microeconomics and the reification of models, move through macroeconomics and then free-market economics which bridges the macro and the micro, to ubiquitous economics, which selectively absorbs and simplifies.

Microeconomics, 'homo economicus' and the reification of models

In mainstream economics individuals are seen as the embodiment of 'utility functions', which describe the things they want and how they prioritize those wants given that there is always scarcity. So, individuals differ in terms of what they desire but how they come to desire different things in the first place is not taken to be the business of economists: individuals are defined as free to make their own decisions based on their desires as expressed in their utility functions. This allows the word 'freedom' to go on holiday in the scientific discourse: 'For markets to be efficient, individuals have to be free to choose so individual freedom and the prosperity of societies go hand in hand', hence the relationship between microeconomics and liberal politics. The economic theory of the market binds individual freedom together with efficiency at the social level. But in the real world, individual choices are never free except, perhaps, if one looks at them at a frozen moment in time divorced from even their most immediate history (which, for instance, advertising is designed to mould just as would any other kind of force).[8]

The implicit assumptions on which *homo economicus* is based are routinely stated in economics textbooks but do not feature as a topic of debate once the modelling is in hand. The model is that the interaction of individuals' utility functions with the resources available to them determines how much demand there will be for different goods and this, when confronted with the supply of goods, sets a price. If the price is high, suppliers will endeavour to supply more goods than if the price is low, and so on. Mathematical analysis shows how, in a 'free market', an equilibrium will be reached that matches demand and supply at a certain price; this is said to be the most efficient way of using and distributing scarce resources in a society while satisfying consumers' wants in the optimum way. No one can predict this price in advance because it is a product of the interacting utility functions of all the individuals in the market, so the market cannot be replaced by planning without abandoning productive efficiency and without imposing central authority on what, under liberal politics, should be individual freedom to choose. Wealth, politics and even religion (of a certain kind) go hand in hand.

This works so long as the idea of 'freedom' is itself given freedom to escape from the societies in which individuals live and so long as moral constraints are found from somewhere else, and so long as the transmission of information is perfect, markets clear and set prices instantaneously, there are no restrictions on competition, market externalities such as pollution and climate heating are ignored, and so long as it is not thought unreasonable that competition collapses markets into monopolies. All these (and more) reservations and qualifications are understood in principle in that part of the institution of economics that is found in universities but, to repeat, analysis of their major implications for 'normal economics' has often been relegated to the status of 'alternative'. Meanwhile, a good part of mainstream economics continues to inform the public domain with physics-like calculations resting on taking the *homo economicus* assumptions for granted.[9]

So long as economists stay within the imagined world of free choice, justifying it by the relative efficiency of markets as contrasted with planned economies, economics has the look of a science like physics. The trouble is that looking like physics is not the same as being like physics, and this applies to big areas of professional physics as much as it applies to economics. For example, string theory is an area of physics supported by an enormously sophisticated and flourishing mathematics, but it does not seem to bear on the world, so, as many physicists argue, it isn't really physics at all.[10] String theory has developed a life of its own and physics, which uses mathematical models intensely, is full of such cases. For instance, before there were any empirical observations bearing on it, the discussion of black holes was cast in the language of discovery, but it was always *mathematical* discovery. But as soon as physics finds ways of bringing empirical evidence to bear on black holes, the models were relegated from reality to the status of tools. The essence of the science is the judgements about what is worth pursuing and for how long; how to build experimental or observational apparatus and, once it is built, how to work out what is a signal and what is noise. It is a matter of how and where to apply the mathematics.[11] In economics, notably microeconomics, the models tend to displace the real world – what we have called reification of models. Perhaps this is because economics is in competition with other social sciences where models are of far less power.

The way modern academic disciplines run themselves allows an appearance of mathematical 'scientificity' to be enough to keep a discipline flourishing whether the mathematical models bear upon the world or not. This is certainly one of the complaints about the flourishing of string theory in physics. Every university wants to recruit string theorists because the people doing the recruiting are string theorists themselves or are impressed by the growth of string theory in competing universities; in an academic market driven by publication, string theory is a winner. In economics the grip of the top journals on what gets published and how those publications are assessed is enormously influential. In the UK this distorting effect on science has been created through the competitive quasi-markets set up by the economics-influenced governments who control the flow of funds; the resulting organizations designed to assess and measure, reminiscent, ironically, of the Soviet Union in terms of the large bureaucratic overhead and flavour of planning, maintain the mathematical game at the heart of a profession like economics while its mathematical virtuosi are often recipients of economics', sometimes controversial, 'Nobel Prizes'.

But now consider how the illusion of the calculations at the heart of the mathematical analysis of markets is supported. Consider, for example, how one has to interpret the world in order to justify altruism or deadly self-sacrifice as a free choice based on self-interest rather than moral integrity.

> There are two ways to get around this without abandoning an assumption of contributors being rational. The first is to assume that they are motivated altruistically (i.e., have other-regarding preferences.) Self-sacrifices are made in order to maximize utility functions that include the benefits to others (e.g., family, compatriots) as an independent variable. Azam (2005) views suicide bombings by terrorists as a form of inter-generational investment. Reducing current consumption to zero by carrying out a suicide bombing is viewed as an extreme form of saving, whose benefit will accrue entirely to the next generation. The second is to assume that the contribution jointly produces private benefits as well as public benefits. Private benefits work as a selective incentive (Olson 1965) when they are contributor-specific. In the case of suicide terrorists, for

> example, possible rewards in the hereafter are often mentioned in the literature (e.g., Berman and Laitin 2008) as such an incentive.
>
> (Shimuzo 2011: 428)

It is not hard to see the unfalsifiable, self-serving work that is going into maintaining a model of self-sacrifice based on the interaction of utility functions. But this work is never going to *explain* why some soldiers or terrorists sacrifice themselves, or some religions, terrorist groups or entire societies, valorize self-sacrifice; it is, at best, going to fit a mathematical prejudice into a pre-existing 'scientific' framework by making it a truism that self-sacrifice is motivated by self-interest. The question of why people are willing to sacrifice themselves has to be answered by other, less mathematical, means. It has to come from the sciences that can explain where utility functions come from. Aside from individuals with suicidal tendencies due to depression or some such, it is a question of culture – once more, the academic economics is downstream of the culture. For instance, martyrs going gladly to their deaths have to sincerely believe that paradise awaits even though no one has ever reported their experience of the afterlife, while in the First World War, British soldiers had to be influenced by something like Rupert Brooke's 'If I should die, think only this of me: That there's some corner of a foreign field, That is for ever England' rather than by something like Wilfred Owen's cynicism about, 'The old Lie: *Dulce et decorum est pro patria mori*'. Culture is upstream of the interpretation of death's utility.[12]

A compelling feature of microeconomics is the 'objective feel' of the numbers that feed into the models. In comparison, a social science like sociology has to 'manufacture' its numbers by conducting surveys; the sophisticated statistics used to analyse survey results often seem to overpower the thin and fragile survey data struggling to support itself with inadequate statistical justifications, whereas prices and volumes of goods seem to be just 'out there' with the ontological status of lengths, masses and temperatures.

Sociology appears to struggle because its starting point is not the individual but the collectivity. The associated methodology requires that analysts must start by immersing themselves in the society or social group they are investigating in order to understand the world of the actors even before they conduct surveys and the like. This need for an initial 'understanding' via immersion makes a subject like sociology open to the *accusation* of 'subjectivity'. Quantitative analysis, should it be thought desirable, begins only after this subjective starting point, whereas economics, with microeconomics as the icon, starts with the culture-free *homo economicus*. We'll argue, later in this chapter, that this is one of mainstream economics' weaknesses rather than a strength; both academic and ubiquitous economics would have more integrity and a more successful approach to correspondence truth if individuals' choices were disconnected from the idea of individual freedom. If individuals' choices are determined by the culture, as they assuredly are, they are heavily determined all the way from place of birth through secondary socialization to continual encounters with advertising. This means that prices are actually downstream of culture too: the ontology of prices, if you are after the truth of the matter, should be seen as just as much associated with the cultural world as the physical world.

Microeconomics appears in Chapter 4 as a science where assumptions, such as freedom of choice, are not questioned in the day-to-day work – they are taken for granted. But this is where mainstream economics starts to violate its preferred value-freedom and instead inserts the value of 'individual freedom' into the embedding society. Even academic economists can fall victim to their own sense of exceptionalism.

Game theory and the virtues of selfishness

Documentary-maker Adam Curtis argues that Cold War game theory, based on the assumption of pure selfishness, worked through the realization that any attack would lead to mutually assured destruction – destruction of oneself as well as the enemy. Micro-economics shares the assumptions of game theory – self-interest drives choice: there is no room for a sentimental morality. Indeed, such a thing is said to be dangerous because the consequent actions are unpredictable whereas the outcome of pure self-interest is predictable and calculable.[13] Curtis shows James Buchanan, originator of 'public choice theory' and one-time President of the Mont Pelerin Society, an important source of market fundamentalism, speaking to camera and explaining that those who have too much zeal for public service are dangerous. They are dangerous because they are bound to support their own theories of how society should work instead of leaving it to the market and allowing self-interest to work its way to the right and predictable conclusion. Curtis intimates that this model of social interaction, perhaps exhibited and supported by the paranoia of founding game theorist, John Nash, feeds back into the national culture.

This kind of thing is hard to document but it is not hard to see that micro-economics and its conjoined twin, game theory, in so far as they do influence the wider culture, would influence it in the direction of individual self-interest being a morally desirable state of society. Micro-economics, combined with the trickle-down theory of wealth at the top turning into wealth for the many, can readily justify 'greed is good'. Instead of citizens experiencing inequality as something they don't want – a disutility – they would be learning to experience it as a good thing for our long-term survival and, at worst, something temporarily unavoidable for utility maximization going forward. How this relates to liberalism is harder to say but it certainly seems to correspond to the notion of free enterprise with just rewards to the successful and, inevitably, fewer or no rewards to the comparatively unsuccessful.[14] That's how mainstream micro-economics would feed into culture via its role in ubiquitous economics and the influence of politicians and that's how the subsequent culture would feed back into academic economics.

Market fundamentalism

The original post-Second World War Austrian theorization of the value of markets had clear but limited aims; the analysis of markets put forward by Friedrich Hayek was intended to show the failure of planned economies and their likely descent into totalitarianism; it was not unreasonable. But Hayek (at least, at the time he wrote *The Road to Serfdom*, 1944) did not believe that, for example, unconstrained markets could deliver moral values.[15] He believed that government regulation would be required to prevent free markets from collapsing into monopolies: monopolies would give rise to unacceptable concentrations of power, destroying markets in their turn, so governments would have to step in to limit them. Furthermore, markets would not take into account what economists call 'externalities', which affect those who take no part in, and gain no benefit from, the transactions but suffer from their undesired consequences. Today's most salient example of involuntary damage caused to distant people by others' exploitation of resources is climate change. In whose market interest is it to limit long-term effects like climate change or even shorter-term effects like the pollution of rivers by the industries lining their banks? Hayek saw most of the problems (climate change was before his time), but successors such as Milton Friedman, defend even monopolies, arguing that, if they arise from markets, they must be the most efficient distributors of goods, while he refers to externalities like climate change and pollution with the warm and soporific term, 'neighbourhood effects'.

Hayek is often cited as the founder of the modern stress on markets, but he was writing out of fear of a return to totalitarianism and realized that markets left to themselves have abundant problems. For what we will call the 'market fundamentalists' who succeeded him, Hayek was too timid in his most famous work. In market fundamentalism, morality, limited to that which is a consequence of free markets, and monopolies, which are also such a consequence, must, by that very fact, be efficient. Liberalism's association with free markets easily becomes libertarianism's anathema of more than a minimal amount of government interference with citizens and their economic activities. Whether 'fundamentalism', with its religious connotations, is the right term can be disputed but given the evidence for the close association of free markets and evangelical Christianity (we will describe the association below), it does not seem entirely inappropriate. Exactly who was responsible for the way the more fundamental version grew out of the less fundamental view of markets, such as is found in the earlier 'Austrian economics' of Hayek and von Mises, is another matter for dispute, though its location was almost certainly 'Chicago School' economics bolstered by an explosion of thinktanks supported by big business as will be described below. Reverting to the vernacular, ex-Chicago School economist Deirdre McCloskey describes her Chicago School colleague Milton Friedman's involvement as follows:

> The famous line from [Milton Friedman's] … essay is half of the sentence – the social responsibility of a corporate executive is to make as much money as possible. But the second half of the sentence goes: while conforming to the ethical and legal rules of society. That part always gets left out. On the other hand … It metastasized in business schools. Business schools went from trying to make managers into professionals – like lawyers, professors, and doctors, who acknowledge a deep ethical commitment to their clients, their students, their patients – into making people with a kind of 'screw you' attitude.
>
> (McCloskey contribution at page 31 of Deneen et al. 2023)

In this quotation, we see a sad reversal of Durkheim's proposed role for the professions. We also see both upward and downward influences: Chicago School economics feeds into the business schools which feed back into the culture of the professions. And, whatever McCloskey says about Friedman, it is hard to see how Friedman could be justifying a moral basis for society and hard to see what he and others with a deep commitment to free markets could say to the business schools to stop the metastasis that is the subject of the complaint: the metastasis, if that's what it was, is born in the logic of the free-market view. Market fundamentalism turns what was a not an unreasonable economic/political analysis into an anti-society doctrine and turned an interesting but low-level and restricted science of human behaviour into an ideology with, as we'll see, religious overtones.[16]

These transformations are only partially supported in the university-based economics community, but the origins are in that community and there has been little effective resistance from that community to the way economics is metamorphosed into ubiquitous economics and interpreted in the political sphere. As will be documented, the ineffectiveness of academic resistance to these interpretations has been a result of feedback from the societies (the left-hand panel of Figure 3.2), either via direct influence on academic appointments, or through the feedback loop from the national culture.

The irony of a macroeconomic technocracy

Macroeconomics is less subject to reification models than microeconomics because the models are intended to predict national economic outturns a year or so ahead and the inevitable failures soon become evident. This means that econometric modellers are continually aware that their models depend on the assumptions on which they are based and the attractions of reification of models are not great nor are the models a great advert for the use of mathematics in social science. One assumption shared by micro- and macroeconomics is the supposed freedom of markets but macroeconomics is confronted by both the downsides of such freedom and the impact of culture on actual economic outturns with, for example, savings rates, wage bargaining and patterns of consumption expenditure all being influenced by events in the world and expectations about the future.[17] In this aspect of its activity, then, macroeconomics approaches the ideals of science with the very disagreement between modellers and their predictions – the very failure to reach the kind of consensus one would hope for in a successful science – indicating a search for correspondence truth associated with moral if not substantive success. (At least this would be the case if the models and the choice of assumptions were not politically motivated, as they often are.)

But there is another aspect of macroeconomic work that raises a different kind of question. Macroeconomists want to make long-term predictions about the evolution of economies at the level of societies. To predict the future of societies as a whole, relating their economic and political choices, is to put forward a 'theory of history'. Marxism turned on a theory of history – that capitalism would inevitably lead to increasing exploitation of the poor by the rich, leading to increasing polarization and immiseration, leading to social revolutions that would usher in the era of communism. Karl Popper, thinking and writing with the same concerns as Hayek and his colleagues, believed that such theories could not be justified; he referred to them as 'historicism'. Popper's *Poverty of Historicism* and his *The Open Society and its Enemies* are critiques of Marxism's theory of history. Popper argued that national historical change was a unique process whereas scientific testing of theories depended on the iteration of the same effect so that they were subject to testing by replication.[18] Popper's critique has been influential, especially among those who favour freedom of action in free markets driven by a fear of Hayek's 'serfdom'.

But once one is convinced that free markets are the optimum way to organize society it is easy to fall back into historicist thinking. One may easily proclaim that if governments are not strictly limited in their scope, they will automatically expand to fill all the economic space there is and give rise to left-wing totalitarian societies – that too is historicism. One may easily believe, given the affinity between certain religious beliefs and free enterprise and that between versions of democracy and free enterprise, that the imposition of Christianity, and of democratic institutions on developing countries will bring free markets and wealth to their citizens. Wars have been justified and fought for these beliefs. But given Popper's critique of historicism, embraced by critics of left-wing autocracy, why should one expect these right- or centre-leaning theories of history to be any more reliable than the Marxism at which the critique was first directed?

They are not reliable, of course, and the empirical evidence does not support them. Wealth can be found under various political regimes and, even in European countries the size of government and the proportion of tax collected is not correlated with the health and growth of GDP; the imposition of political regimes favouring small government and low taxation has not led to wealth in all countries even though some, such as the United States, are wealthy (though, in this case, given high levels of US government-sponsored military spending supporting industry, exactly how this illustrates the positive effect of free markets is not so clear). On the other hand, some countries with relatively high taxes supporting proactive governments are wealthy and many countries on which democratic institutions have been imposed from outside are neither wealthy nor democratic, at least, not for very long.

The irony we have pointed out above is that, in spite of these weaknesses in economics as a science, it is treated as the most technocratic science of all. It is, to repeat, the one science where the scientists, the politicians and the public are ready to endorse a move from science to policy without anything in between.

The still deeper irony is that those who believe in the free market as the only way to run a society claim that markets are the way to avoid technocracy because markets take decision-making away from any centralized power, including scientists, and distribute it among the individual people. Yet even as they generalize the claim about the economic superiority of markets and their link with democratic governments, they are supporting technocracy by acting out a scientific theory of economic and political development quintessentially linked to policy.

Ubiquitous economics

Why economics has an incarnation as a ubiquitous science is an interesting question. Of course, it is, in one way, a rejection of elite science, like taking social media seriously in matters of science and preferring conspiracy theories. In this case, however, the rejection of the actual authority of science is enacted by the endorsement of what is thought to be science's authority. It may be that citizens feel they have a grasp of it because they deal with economic issues every day of their lives and it may be that politicians too think they have gained economic expertise through their experience or their education, but it is also the case that there is some benefit to politicians in enhancing the apparent scientific power of economics because it enables them to seem to delegate the responsibility for unpleasant political decisions to science. Presenting the science of economics as belonging to everybody makes the chosen policies seem democratic and fair. If the science of economics says that wages in the public service must be capped to keep down inflation while taxes on the rich must be decreased to encourage enterprise, then 'TINA' – There Is No Alternative – and the majority agrees and will express that agreement at the ballot box.[19]

An earlier example of this way of deploying science is South African President, Thabo Mbeki's, 1999–2008, denial of the safety and efficaciousness of anti-retroviral drugs, even refusing them to pregnant women and newborns (see Chapter 5). Mbeki claimed (to the members of South Africa's second parliamentary chamber) that his decision rested on the scientific controversy that they would find if they researched the internet. His decision, he claimed, was precipitated by the science and the parliamentarians did not need to be specialists to understand the controversy, they could manage with just a little reading and their existing wisdom. But the controversy on the internet did not reflect an extant controversy in the scientific community; the scientific consensus was that the drugs were safe and efficacious. The parliamentarians reading the internet were *not* equipped to know this.

TINA in the case of economics is equal and opposite to the Mbeki case. The professional world of economics, at least the contemporary world of academic economics, *is* riven by disagreement, but it is presented as uniform and consensual to the extent that it determines policies rather than simply offering debatable advice. Mbeki, on the other hand, was presenting a scientific consensus as a disputed domain.

Ubiquitous economics is the economics shared and discussed as colleague-to-colleague by politicians and the citizen. Ubiquitous economics is not *popular economics* in the way that, say, popular physics is a watered down, journalistic, version of physics. Popular physics is, intended to provide an engaging and entertaining diversion. It is very much part of the 'contingent forum', where the 'constitutive forum' is where knowledge is generally thought of as being made – laboratories and publications; the contingent forum is where knowledge is talked about – the coffee bar and the newspapers. A ubiquitous science such as ubiquitous economics melds the constitutive and contingent forums.[20] Popular physics is not to be made use of by the public in their day-to-day technical decision-making whereas ubiquitous economics is presented by politicians and the media in the argumentative forms appropriate to the constitutive forum, if not the technical detail. In ubiquitous science, public reasoning is invited into the scientific process. Ubiquitous economics is offered and consumed as a technical subject regarding which politicians, together with voters, can debate and reach consensus on how to run the country: the distinction between the constitutive and contingent forums is dissolved; the distinction between the inside and the outside of Figure 1.2 disappears. To repeat, the unusual thing about ubiquitous economics is that citizens and politicians are not treated as though they were limited, as in the case of other sciences, to making judgements by assessing and ranking the trustworthiness of the institutions that produce the advice, they are treated as though they can do the economics for themselves, even if they aren't actually doing any technical economics with all its disagreements. Going back to Table 4.2, the citizens are not limited to the upstream or downstream portals, they are right there at the research front or, at worst, like expert peer-reviewers, considering and judging how the economics of their society works and voting accordingly. Or, in other words, economics is treated as a ubiquitous expertise.

As intimated, that degree of participation is more typical of the science that informs conspiracy theories even if in that case it is opposed to elite science rather than supporting it. For example, many citizens consider themselves able to work out that a child's immune system cannot possibly cope with three vaccines injected at once and then argue that this is empirically proved by the autism that follows such injections (as in the revolt against case of measles, mumps and rubella vaccine).

Of all people, Friedrich Hayek seemed well able to understand the dangers of ubiquitous economics. In 1974, Hayek was the recipient of the economics prize 'in memory of Alfred Nobel'. In his banquet speech he said:

> [I]f I had been consulted whether to establish a Nobel Prize in economics, I should have decidedly advised against it …
>
> [There are two reasons, the second of which is the most important] It is that the Nobel Prize confers on an individual an authority which in economics no man ought to possess.
>
> This does not matter in the natural sciences. Here the influence exercised by an individual is chiefly an influence on his fellow experts; and they will soon cut him down to size if he exceeds his competence.
>
> But the influence of the economist that mainly matters is an influence over laymen: politicians, journalists, civil servants and the public generally.[21]

Hayek was conscious of the strange status of ubiquitous economics described in this chapter.

The post-war history of economics and how it came to be the way it is

How did we get here? We begin by describing the development of economics and the rise of ubiquitous economics in the post-war period, paying particular attention to the United States and UK. In the West, there were strong societal influences on economics favouring free markets over central control, but a level of healthy disagreement was maintained in the academic world where the assumptions could still be questioned. In contrast, ubiquitous economics, where these disagreements were not evident, followed the free-market argument as it was developed in the Chicago School and took free markets to support individual liberty and freedom in general. Here, then, economics itself could be said to be influenced by ubiquitous economics rather than ubiquitous economics simply being an enchanted version of academic economics. It is as though Trump's claims about alternative treatments for COVID, accepted by his followers, were to feed into the medical science.[22]

We have distinguished between 'academic', 'mainstream', 'alternative', free-market and 'ubiquitous' economics. To repeat, academic economics refers to all the different fields and subfields of economics that can be found within universities. Mainstream economics is a branch of academic economics that has gained notable power within the academic domain partly as a result of outside influences on the academic subject. Without those outside influences free-market economics may not have gained the salience it has and mainstream economics, with which it is more compatible than other kinds, may well not have been so dominant in its relationship with alternative economics. Without those outside influences, free-market economics' influence on academic economics, and its impact on ubiquitous economics would almost certainly have been far less. We illustrate what we have said in Figure 6.1.

Figure 6.1 is an *aide memoire* for the place of ubiquitous economics in the fractal model. The 'economists' oval towards the bottom left represents mainstream academic economics but it includes two subdomains. The small dark disk with a black dashed outline represents heterodox economics. The white ragged star represents free-market economics. These have different levels of influence on the mainstream. A large white ragged star has been added to the ubiquitous expertise oval and labelled 'POLITICS'. This is meant to represent the role of political discourse in public debate and the way it helps form the substance of ubiquitous expertise with politicians having well-funded and widely broadcast outlets for proselytizing about their preferred public values; a significant component of this influence is ubiquitous economics. The curved arrow indicates the strong influence of free-market economics on ubiquitous economics and on the public as a whole through the work of think tanks and lobby groups.[23]

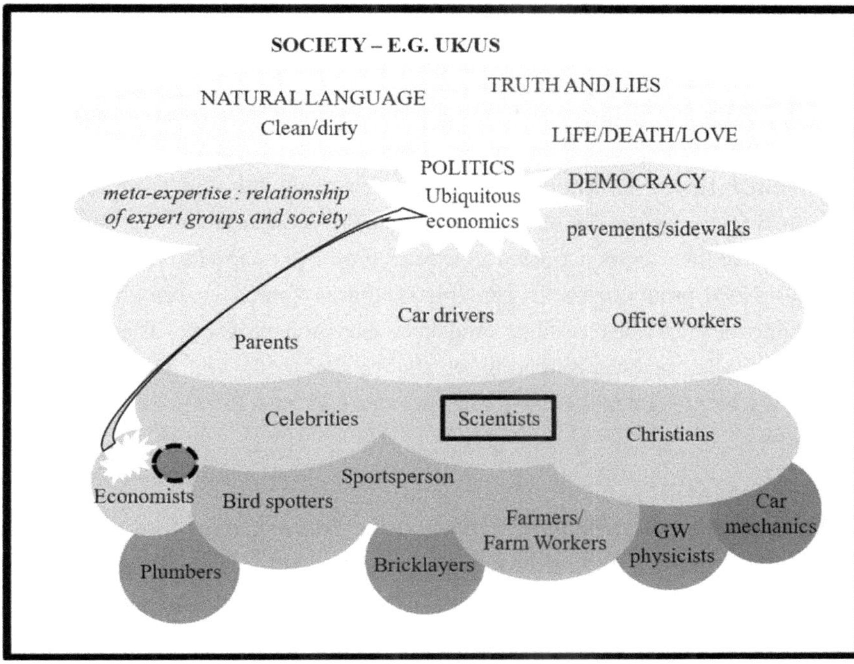

Figure 6.1 Economics and ubiquitous economics represented in the fractal model (created by Harry Collins).

Popular culture and political think tanks

Naomi Oreskes and Erik Conway provide a detailed picture of outside influences on US economics and their work has been a vital input to this chapter. Published in 2023, and informatively entitled, *The Big Myth: How American Business Taught us to Loathe Government and Love the Free Market*, the book explores the foundations of twentieth-century culture in the United States. The key move is what came to be called the 'indivisibility thesis', in which individual freedom and laissez-faire capitalism are said to bring mutually reinforcing benefits to society. Oreskes and Conway show that industry bodies, such as the National Association of Manufacturers (NAM), at one time used this indivisibility thesis:

> to argue against political reforms that today we take for granted, such as laws limiting child labor, establishing workmen's compensation, and creating the federal income tax. In the 1930s, they aligned themselves with the electricity industry and used the thesis to argue against the Rural Electrification

> Administration, the Tennessee Valley Authority, and other elements of the New Deal. …
>
> In the 1930s, NAM [organised] a multi-million-dollar propaganda campaign to convince Americans … that the real causes of the Great Depression were unionized labor's unreasonable demands, coupled with the government's excessive interference in business affairs and federal taxation that starved industry of the monies it needed to expand productive capacity. Well into the 1940s, NAM produced books, pamphlets, radio programs, lecture series, and documentary and feature films (and later television programs) designed to influence what newspapers had to say about the economy and American life, what teachers taught in the classroom, and, above all, what the American people believed.
>
> (Oreskes and Conway 2023: 6–7)

To give one example, *The Little House on the Prairie* series of books showed how settlers extracted wealth from an 'empty' land, and though it was, purportedly, the true story of a settler, it had been edited into a tale of self-sufficiency and the superfluousness of government by the settler's daughter, a libertarian journalist.[24] In a similar way, the TV series *General Electric Theatre*, sponsored by the electricity provider and manufacturer, was designed to show how big business sponsored a good life for those with the strength to be self-sufficient. Through these and many other examples, Oreskes and Conway show that the manufacturers established a 'tripod of freedom', claiming that America was founded on the interdependent principles of representative democracy, political freedom and free enterprise. Of course, the third element had nothing to do with the American Constitution but was shoehorned in by the manufacturers who, we presume, saw the resonance between the individual liberty of political freedom and the free choice of the economic market.

Oreskes and Conway also show that religious organizations played a significant role in spreading and legitimizing these ideas.

> Through the efforts of Spiritual Mobilization, *Christian Economics*, and *Christianity Today* – reinforced by FEE [The Foundation for Economic Education] – the captains of American industry had found a way to turn Protestant theology on its head, from embracing the poor to celebrating the rich. They channelled corporate profits to promote as benevolent an economic system that often had been anything but, and to vilify government efforts to address that system's failures. Crucially, by embedding capitalist propaganda

> into mainstream religion, they ensured that millions of Americans heard their message weekly, in church, from their ministers and lay leaders – which is to say, from people they thought they could trust. … economic historian Daniel Stedman Jones refers to neoliberalism as 'faith-based policy', because it has often proved impervious to evidence. In its Christian libertarian expression, neoliberalism was not just metaphorically faith based. It was literally so.
>
> (Oreskes and Conway 2023: 313–14)

So here we see how a certain version of economics feeds into the cultural life of American citizens. But how did that version of economics become so dominant? It appears to have had quite a lot to do with the influence of the same American manufacturers: not only were they spreading the view to the public and politicians, they were also influencing the way academic economics was practised. The same business groups that promoted free-market values in popular culture also helped Ludwig von Mises and Friedrich Hayek, analysts and supporters of free markets, to get academic jobs in America. They were particularly influential in what became known as the 'Chicago School'. Business also founded and funded think tanks dedicated to promoting their views. Without this support, it is entirely possible that von Mises and Hayek would have disappeared into obscurity as they had no great status as economists in their native Austria. For example, Hayek was not accepted into Chicago's School of Economics, but had to be located, initially, in a law programme at Chicago and von Mises's New York post was funded privately.[25]

We don't have as much evidence of the direct influence of big business on academic economics as we have of its influence on popular culture but think tanks look both ways, offering promising positions to academics who fit their projects so making their kind of economics more attractive in terms of job security, as well as influencing political and public culture. Furthermore, as we have argued, there is a feedback loop between culture and economics mediated through the model of the world which potential economics students are endowed with in their socialization.

Business groups also helped create America's first libertarian think tank, the Foundation for Economic Education (FEE). It promulgated pro-market, anti-government ideology. And they helped create the Hayek-aligned Mont Pelerin Society. They also helped Milton Friedman, another Chicago School economist, to write *Capitalism and Freedom* (Friedman 1962) and to be awarded the National Medal of Freedom by Ronald Reagan in 1988. Their various programmes of support came under the heading of the 'Free Market Project'. George Stigler, yet another Chicago School economist, would edit a version of Adam Smith's *Wealth of Nations* removing Smith's caveats about free markets such as the need for bank regulation and wage regulation and the need for taxation to support roads and bridges and similar public goods, rather as the caveats in Hayek's *Road to Serfdom* would be lost to sight in later versions of free-market doctrine. Many other market-favouring think tanks were founded and funded by big business and they are still prominent employers for liberal-minded academics.[26]

Similar events can also be seen in the UK, with Friedrich Hayek once again playing a key role according to documentary-maker, Adam Curtis.[27] In the UK, the catalyst for ubiquitous economics was Anthony Fisher, a right-leaning entrepreneur who would bring factory farming to the UK with his firm, Buxted Chickens, also making him a millionaire. Fisher had read an abridged version of *The Road to Serfdom* in a 1945 edition of *Reader's Digest* and had become a convert to Hayek's free-market ideas.[28] Hayek's book argued that government control of the economy would lead to communism. Fisher became convinced that the Keynes-inspired state planning that gripped governments after the Second World War would lead to totalitarianism. In the 1950s, he visited Hayek, then teaching at the London School of Economics, and asked for advice on how he could counter the Keynesian orthodoxy – should he go into politics? According to Curtis, Hayek told Fisher:

> Politicians are trapped by the prevailing public opinion. Instead, Hayek said, Fisher should try and do something much more ambitious – he should try and change the very way politicians think – and the way to do that was to alter the climate of opinion that surrounded the political class. Fisher wrote down what Hayek said to him.
>
> *He explained his view that the decisive influence in the battle of ideas and policy was wielded by intellectuals whom he characterised as the 'second-hand dealers in ideas'.*
>
> Hayek told Fisher to set up what he called a 'scholarly institute' that would operate as a dealer in second-hand ideas. Its sole aim should be to persuade journalists and opinion-formers that state planning was leading to a totalitarian nightmare, and that the only way to rescue Britain was by bringing back the free market. If they did this successfully – that would put pressure on the politicians, and Fisher would change the course of history.

Hayek was proposing that Fisher change popular culture which could then influence economics in the free-market direction.[29]

Fisher took up the challenge with gusto, setting up the 'Institute for Economic Affairs', a right-wing think tank that continues to be influential to this day, and, according to Curtis, he founded around 150 other think tanks. Whether there is a direct connection to the IEA or not, we know that Margaret Thatcher is said to have carried a copy of Hayek's *Constitution of Liberty* in her handbag, and she is notorious for her claim that there is no such thing as society, only individuals and families. And Thatcher would go on to form an alliance with Ronald Reagan, who was another convert to Milton Friedman's brand of free-market economics, which had become increasingly influential in political circles during the financial crises of the 1970s. It was Ronald Reagan, in 1981, picking up the libertarian/economic mantra of small government, who said:

> In this present crisis, government is not the solution to our problem; government is the problem. From time to time we've been tempted to believe that society has become too complex to be managed by self-rule, that government by an elite group is superior to government for, by, and of the people.

The increasing influence of Friedman's ideas saw Hayek gradually displaced from the neoliberal pantheon. Free-market ideas grew ever more strident, and the political centre-ground tracked steadily to the right. Everyday discourse also followed these changes, so that what we have called ubiquitous economics came to be characterized by three ideas: an unswerving commitment to the idea that free markets are always preferable to state provision; that governments are always the problem and never the solution; and that preserving the autonomy of the individual is the highest priority in any political system or debate.[30]

Ironically, then, the outcome of Hayek's advice to Fisher is the creation of the very thing he appeared to worry about when he was awarded the economics prize in memory of Alfred Nobel in 1974. Remember, in his banquet speech, quoted more fully above (p127), he said, remarking that he would not have been in favour of establishing an economics Nobel if he had been asked:

> … the influence of the economist that mainly matters is an influence over laymen: politicians, journalists, civil servants and the public generally.

As a result of the think tanks set up with his encouragement, Hayek has perhaps done more than anyone to enable the spread of 'second-hand ideas', with the rise of ubiquitous economics and its influence on politics and everyday life a visible testimony to the influence of economics.

There can be no doubt that free-market economics has been supported by big business to a quite unusual extent for a social science. We are not and should not be surprised that space science is supported by big business, and that computer science is supported by big business and that aviation and military science are supported by big business. We don't think this necessarily causes a problem for the science that is produced – we don't think commercial funding necessarily gives rise to bad science except in the sense that the more arms are manufactured the more can it lead to war or, on a smaller scale in the case of organizations like the National Rifle Association, to civilian deaths. The question we have to deal with is whether support by big business necessarily distorted and distorts academic economics or did it just give rise to a richer and more powerful economic science, the social science equivalent to the products of Silicon Valley?

So, what work does the account of the historical influence of big business on academic economics do if the origins of a science are not necessarily a critique of its substance? The answer lies in distinguishing between upstream provision of resources and interference at the research front. The difference is between, say, support for biology in general, which is benign, and support for Lysenko's version of biology because it fits Soviet national values. If the support is provided in such a way that it distorts research front science in favour of Lysenko, then it is a problem for the substance. In the case of economics there is an imbalance in the sources of economic ideas caused by the imbalance in the sources of funding. This limits and distorts the natural diversity found in university-based economics, with its mainstream and heterodox streams, helping to limit the influence of the heterodox and favouring free-market ideas. The balance is distorted further when things are simplified as they are bound to be in ubiquitous economics (distance lends enchantment) and the problem is reinforced by feedback from the societal culture.

In Chapter 5, we distinguished between demographic and topic-related diversity. Social justice is achieved through demographic diversity, which ought to reflect the diversity of types of person in the population in the high-prestige profession of science just as we would think it just that the distribution of those types was faithfully mirrored among the rich and the poor. But this concept has no application here. Topic-related diversity does have an application, however. It turns out that the way the general population influences much of economics, mostly inadvertently from upstream of the research, is to fund think tanks via funds from the businesses from which it purchases products. We know the sole aim of those think tanks was in the words of Hayek: '*to persuade journalists and opinion-formers that state planning was leading to a totalitarian nightmare, and that the only way to rescue Britain was by bringing back the free market*'. We know, from Oreskes and Conway's research, that the aim of the think tanks they describe was similar. The aim was to create economic institutions of a type that did not reflect even the minimal diversity found in universities: the new institutions would favour free markets and small government, not Keynesianism or its counterparts. And this influence didn't come via the normal consensus building of the closed expert community – the boundaries of the core-set were not impervious but open to a particular group of outsiders with an interest in developing an economic monoculture that would benefit them and using it to influence government policy as well as recruitment inside the academe. And they succeeded!

The procedures of science, such as peer review and blind testing, that Wave Two showed are not as effective as Wave One took them to be, are designed to eliminate unconscious influence. So, it is easy to imagine that the scientists subject to the influence might be sincere, perhaps not even noticing that the new institutions hired only those likely to be in favour of free markets. Or remember the sincere but misplaced decisions of the scientists described in Oreskes and Conway's earlier book, *Merchants of Doubt* (2010). Influenced, perhaps, by the paranoid atmosphere of the Cold War, and the McCarthy witch-hunts, remembering that a communist Cuba was on the doorstep of the United States, they distorted their own scientific findings so as not to be seen to be supporting government interventions in, for example, restraints on tobacco advertising. The ubiquitous economics influenced by big business is clearly compatible with political sincerity in its practitioners though the case is not easy to make in its instigators.

Interim conclusion

This analysis of economics has been intended to serve various very different purposes. The most important purpose is to help to understand our claim that science can be a foundation for truth in society – it can fulfil the role of a low turtle because it itself rests on a turtle very near the bottom – a moral turtle. Ironically, the very impact of economics as a science on societal culture shows that the right-hand panel of Figure 3.2 is not just fantasy: sciences can influence societies. Unfortunately, the influence we have used to exemplify this point is just the opposite of the one we want to bring about. We have also looked at the left-hand panel of the figure and how it has worked in the case of this science. In doing so, we have we posed a problem for the democratizers and the analysts of scientific values: do they applaud this particular influence since it satisfies what they want to see in terms of the relationship between science and the public, ubiquitous economics fulfilling the democratizers ambitions like a dream of co-production, or do they need to find ways to distinguish the values they want from those they don't want. If the latter, we have offered them a solution in terms of virtual diversity described in Chapter 5. This chapter has shown that not just any science can serve as a fulcrum science, a science on which a truth-valuing society can be founded. The sciences we need for this purpose are those that value the search for correspondence truth above every other kind of success. Parts of economics may be one such but some of it is evidently not.

Can economics be a fulcrum science and what would it look like in a veritocracy?

What would economics have to look like if it were to inform public debate in a manner that supported veritocracy? The key to this would be a new relationship between the complexity and debate that exists within academic economics and the representation of that to the public via *popular* (not ubiquitous) economics. Specifically, as with other sciences where there is significant controversy and uncertainty, these complexities would need to be acknowledged and discussed as part of the policy debate with judgements making clear how the balance between competing 'upside' and 'downside' risks has been reached.[31]

Economics, if it is to be a fulcrum science, would have to present itself for what it is in academe, another social science doing its best to advise governments while recognizing the profound disputes in its core assumptions, activities and conclusions, recognizing and avoiding outside influences so as to maintain the aspiration of generating truth, difficult though it is in such a human-centred science. It would have to try to stop itself being distorted, and then transformed, into the kind of ubiquitous economics that has lent itself so readily to technocracy. Economics' advice on the future of society would then be listed as something that needs to be considered when making policy choices, but which does not, by itself, determine what those choices should be – there can be no TINA.[32]

Academic economics, instead of being so strongly influenced by market models, should allow itself to be influenced by, or even become integrated with, the other culture-exploring social sciences, resisting the idea, in respect of insiders and outsiders, that its mathematical apparatus signifies that it is the only true social science. If economics and its utility functions are downstream of culture and economics is sincere about maximizing the aggregate utility of citizens, then it must bring into salience how culture and other influences on citizens' choices interact before we get to prices. This may already be a part of macroeconomics, but it is not salient in the public domain. The figures below show what individuals are and how they would have to be thought about in economics if economics is to become a fulcrum science.

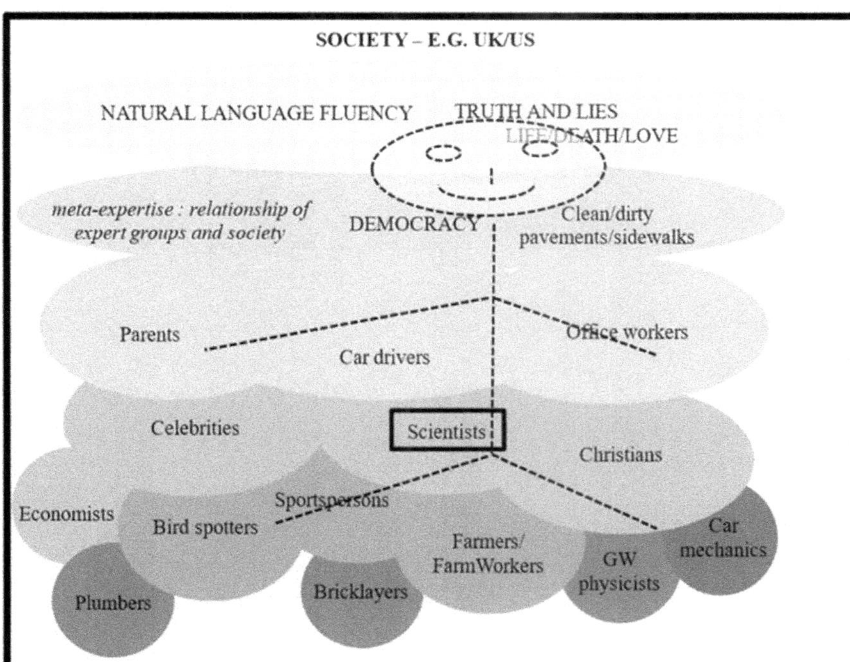

Figure 6.2 The individual is the collection of the expertises into which they are socialized, usually without conscious choice, through primary and secondary socialization.

As discussed, and illustrated earlier (e.g. Figures 3.1, 3.2), society can be thought of in terms of the analogy of a fractal made up of domains of expertise. Figure 6.2 uses the fractal model to reinforce the idea that the individual is not free but is better represented by a sociology of knowledge model. The individual is comprised of the set of domains in which they have been socialized. All citizens with a standard upbringing in a society will share the set of ubiquitous expertises and aspirations that define the society, which they did not choose and over which they continue to have very little choice even in the course of secondary socialization. The individual does go on to choose their secondary socialization, at least to some extent: they 'choose', in so far as they have a choice, which educational and work or professional paths they will take in life. The individual represented in Figure 6.2 is, among other things, a parent, a car-driver, an office worker, a scientist – in particular a GW physicist, a bird spotter, a sportsperson and a Christian. But once these large-scale choices are made, the details of where they take them are unknown. Each subgroup is its own little society, and each provides its own determining features.

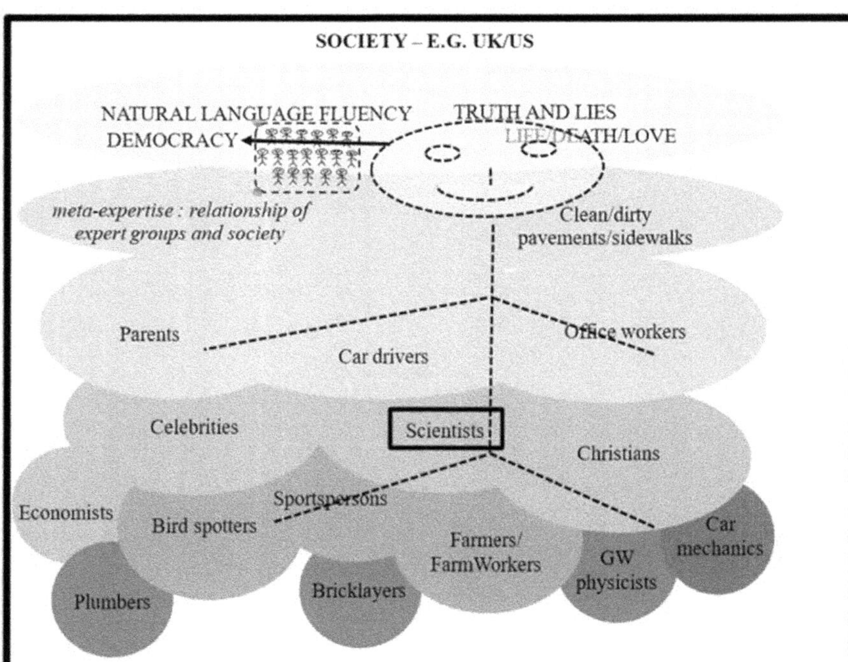

Figure 6.3 Another constraint on individuals' freedoms – their responsibilities to others.

Figure 6.3 illustrates another constraint on the libertarian idea of the free individual taken from the theory of liberalism itself.[33] The claim that 'Your freedom stops at the end of my nose' was originally expressed by John Finch.[34] An individual's freedom cannot constrain that of other individuals or there can be no freedom in society and the point is illustrated with the collection of others that the virtuous individual must visualize inhabiting the space between their choices and the moral domain. That is why we allow governments to constrain our freedom in respect of the side of road we drive on and constrain us to drive at uniformly agreed maximum speeds in different locations, and to make it illegal to drive while under the influence of drugs. Those restrictions enter the moral codes of societies and are mostly incorporated into the contents of primary socialization, though some aspects, such as the way 'my' refusal to vaccinate myself impacts on your freedom not to be infected by others, are not universally understood.

Note that to understand the impact of something like vaccination on others' freedom takes science: we need science to know where, in the manner of speaking, the end of your nose is located. We need science to tell us about the effect of speed limits, and the effect of drugs, and their consequences for driving related deaths. We would need science if we wanted to discharge our slogans about, say, gun control and its effects. And most pertinently for the argument of this book, we need science if we are to understand the externalities of climate change and vaccination hesitancy. The scientific consensus is that a changing climate and refusal of vaccination affects other's freedom so cannot be justified simply by the liberal demand for freedom of choice. Without science, liberalism cannot work as a political philosophy.

Veritocratic governments would also have to understand culture so that they could prepare manifestos that would be attractive to citizens – that would reflect what citizens want. But, as we will argue in Chapter 7, governments cannot avoid the responsibility of affecting culture too – we *do* want a 'nanny state', for without a nanny or equivalent the citizens would not acquire the civic education to make choices: they would be self-interested animals, with only the fittest surviving, as extreme versions of ubiquitous economics prefer to see them.[35]

The government in most countries provides the educational infrastructure to create new citizens and it cannot escape the responsibility of deciding on the substance of government-sponsored civic education. This education includes the moral code, the code of honour, the law, the constitution, elements of science and health, an understanding of the economy and an understanding of the interaction of society and the individual. With the free rein the economics of free markets has been given over the last half-century, especially in the United States and UK, founded in the political philosophy that preceded it, citizens have not been soundly educated into the sources of their choices. They have been led to think they are free, whereas nearly every choice they make is caused by social influences of various kinds and must be constrained by governments if they are not to be subject to the causes of all manner of early death and injury, not to mention the absence of infrastructure needed for economic interaction.

We have argued that the rhetoric of freedom, encouraged by the way particular features of academic economics are distorted and moulded as they metamorphose into ubiquitous economics, is one of the things that stops us understanding the limits of our choices set within the 'organic face' of society – the face that makes it one kind of society rather than another. Another thing is the very mechanism of elections in democratic societies, which constitutes the 'enumerative face'. At election time, votes are counted allowing the citizens (or at least some them) to choose between a very narrow range of options concerning details of the way the society is run – for example, in the United States and UK at this moment in history, we cannot choose to reinstitute witch trials, we cannot choose to institute farm collectivization, and we cannot choose to reinstate a nationalized railway system – because these are not features of the organic face of our societies, the latter having ceased to be a possibility only in the last half-century. But, on the other hand, the outcome of voting – which may well be in the statistical noise of vote counting, and certainly has been in some recent elections – is often described as, 'the will of the people', when it is no such thing, endowing it with a significance quite unrelated to the significance of the causal sequences that made it turn out one way rather than another.[36] At election time there is an obsessive concentration on small differences between parties, agencies using all their powers to bring about what is often the marginal difference between numbers of votes that will determine who rules. The insignificance of the causal sequences usually matches the insignificance of the difference between party policies – a set of small choices set within a much more significant set of determinants of choice and action, formed by the culture and which usually remains stable through many generations. This difference between the organic face and the enumerative face encourages the view that voters live in a world of freedom of choice. But bear in mind what it is we are arguing here: that small differences in policy between parties can mask huge differences in integrity and respect for truth between parties, which could easily lead to vast difference in the political future of the countries involved, vast difference which might make themselves felt very soon in America (more especially since the re-election of Trump).

A cultural economics

We can imagine macroeconomic modelling becoming a fulcrum science quite easily if its practitioners were to express the judgements and uncertainties that underpin their forecasts more clearly.[37] Macroeconomic modelling knows that it is always going to encounter disagreement and knows that its assumptions are crucial and various; it belongs in the top section of Figure 4.1. Macroeconomic modelling is something that responsible governments have to engage in when making policy even though they know the advice is unreliable: they have to accept that, unreliable though it is, it would still be the best bet given only a greater separation from politics. All this could be explained to the public, crucially including the fallibility. As for the rest of economics, it would have to engage with the sciences of culture and engage with their methods too. These methods involve an initial understanding of society followed by surveys of likely actions among the citizenry and the still less quantitative methods of ethnographers and anthropologists.

One such method is consultation with 'native informants', usually a term restricted to anthropology. The authors have generalized this into a sociological method which they call a 'probe'. Where a society is uniform in its actions (as with the small tribes studied by anthropologists), the information provided by a native informant will be reliable because everyone in the society would give the same answer (or to put it more technically, the standard error of any survey would be zero). If the reaction is uniform, one needs only minimal, haphazardly gathered, indicators. An extreme example of uniformity within an entire complex society is the position of the verb in the English sentence. To know this all one needs is to listen to one person speaking, or perhaps a handful in case the first person encountered was mad.[38]

We pick on the probe among the many detailed methods of social investigation because it is also found in economic activity, such as the wise businessperson taking advice from market makers in circumstances where markets are likely act uniformly. An intriguing example emerges from the story of the 2008 financial crash consequent on the collapse of the sub-prime mortgage derivative market. Once more, no model could have predicted the date of the crash as no model can predict the date of the collapse of any market bubble – there's too much complexity. But one account suggests that it was clear that there was a collapse on the way with the same kind of certainty as climate change is predictable, and in this case a 'probe' was enough to see what was going to happen. Michael Lewis's book *The Big Short*, recounts how a small group of investors bet on the imminent collapse by 'shorting' the sub-prime derivatives, but the strain grew as their investment looked more and more unwise while the market continued to be buoyant. Nervous, the investors, effectively, conducted a probe. It is said that based on rumours, they travelled to a Southern location where they spoke to a stripper and her mortgage provider. The stripper, not a rich woman, owned five houses with 100 per cent mortgages obtained without collateral; the mortgage agent was happily taking commissions with his investments guaranteed by insurance policies. It was easy to see that such an arrangement could not survive even a relatively small number of defaults. It is true that the moment and proximate cause of the coming collapse could not be predicted and, up to that moment, everyone else was making money, but it was clear from the probe that the collapse was going to happen before long. And those who bet against the bubble did keep their nerve and, helped by this piece of evidence, made a fortune.

This is rather like predicting the climate: one can do well out of inaction until the climate collapses and the sequence of collapse cannot be exactly predicted, but that the disaster will happen in the not-too-distant future is clear. Going back to the sub-prime mortgages and the stripper, it is an example of what we are calling 'cultural economics': a simple piece of observational social science and some of the judgement that, we have argued, is the essence of social science, with the scientistic mathematics, merely a useful resource. These are some of the ways economics could become a fulcrum science.

Ironically, we are ourselves promulgating what might be thought of as a ubiquitous science, namely the sociology of knowledge view of the world as represented by the fractal model and the stick figure individual constructed on its foundation. This is the opposite of the idea of free individuals interacting to form a market because in this model individuals are not free. How can we defend this opposite view as something that should be available to the public? The answer is that in this case, if citizens were ever exposed to it, they really could check it for themselves. It is not like citizens being expected to understand that lower taxes along with the idea of trickle-down lead to greater wealth for all: that's something that takes a huge amount of scientific work to establish and something that is subject to a huge amount of dispute. But any citizen can immediately understand that their choice of spoken language is not free: if they are born and brought up in an English-speaking country they can't suddenly choose to speak in Chinese or Japanese, their choices are limited by their social group; likewise, for most of their supposedly 'free' choices. This is readily visible to anyone who travels to other countries where things are done very differently. And they can also observe with their own eyes that their television and any other media they consume are full of advertisements and that these advertisements are considered so important in the way markets for various goods turn out that tobacco or fossil fuel businesses are willing to distort scientific evidence to create mock controversies about the safety of their products in order to continue advertising even when it harm has been established. These things are obvious to citizens and that is even before we get to 'subtleties' such as the curtailment of others' freedom that comes with a refusal to vaccinate. Indeed, the bones of this argument are so clear and evident that 'science' is too grand a term: it ought to be just common sense. The strange thing is that somehow the market fundamentalists have managed to obscure these social things while managing to promulgate the science of markets as true common sense.

Summary and conclusion

We have used economics as a case study to explore and illustrate the various claims made in this book. Economics has features that bring out both the left-hand and right-hand panels of Figure 3.2. Though as an academic discipline it is like most social sciences, exhibiting substantial disagreements of principle as well as fact, both the contents and the balance of the disagreements have been affected by the outside influence of big business and politics – the left-hand panel of which ought to make a perfect case study for what we have called the democratizers, and also for aspects of 'scientific value analysis'. The right-hand panel bears on our major thesis but in a backhanded way in the case of economics. We can see how the development of economics in its ubiquitous science incarnation shows the power of individual sciences to affect society, but as far as our major thesis is concerned, it affects it in the opposite direction to that we claim can be the saviour of democratic society. Economics is not a bastion of truth but exhibits the reification of models with too little reference to reality and ubiquitous economics is a bastion of free-market fundamentalism and historicism.

We have leaned heavily on internal critiques of economics already found in academe, if somewhat impotent to change it, as their usual sobriquet, 'alternative economics', and the contrast with 'mainstream' make clear, and we have leaned heavily on historical sources to help us see how it is that the balance between mainstream and alternative has taken the shape it has. This is, on the one hand, because of the left-hand panel type influence on the substance of academic economics and the right-hand panel type influence on think tanks and their feedback effects on the academe. There is also another feedback effect: economics is the most mathematical of the social sciences and to those who do not understand science deeply but accept the self-descriptions of physicists which, perforce, we lived with for most of the twentieth century because there were no other descriptions of science available, this makes it seem the only true social science. Governments, needing a justification for the economic policies that they have no choice but to make are delighted to grasp hold of this idea, especially in countries where conservatism is strong, since it also justifies policies that give free rein to the market ideas that are found in the ubiquitous incarnation of the subject and which are favoured by major financial supporters of their political parties who are, at the same time, supporting free-market ideas in the heartlands of economics in the ways described above. It must be hard for many economists to resist the accolade of being the only true social science when mathematical models are taken to be the criterion and mathematical virtuosity is the route to success within the discipline.

In the UK, in a bizarre caricature of this kind of thinking, in 1983, Margaret Thatcher's minister Keith Joseph insisted on renaming the UK's principal funding agency for the social sciences, the *Social Science Research Council,* the *Economic and Social Research Council* because he thought economics was the only social science that was a real science with falsifiable predictions.[39] The downside of this kind of interaction is that much of economics has come to prefer the perfection of mathematical models to the assiduous pursuit of correspondence truth, a pursuit which, as we have explained, would require economics to take on a much more cultural aspect. Unfortunately, this means that economics, other than that part of the subject where the disputes are still at play, cannot fulfil the role of a fulcrum science – a science where the pursuit of correspondence truth and its correlate, moral truth, can form a foundation for a truthful society.

In a sense we are returning to an old sociological topic which is the tension between science and its applications. A standard presentation of the problem is

Cotgrove and Box's *Scientists Industry and Society*, dating back to 1970, which sets out the tension. But we can see it in front of our own eyes every day if we compare physics and the science of artificial intelligence. Physicists' worst fear is announcing something that turns out not to be true; the motto of Silicon Valley is 'move fast and break things', and its irresponsibility in terms of truth is evident in the continuous and ludicrous claims to have built programs that pass the Turing Test. In Chapter 2, we argue that the 'least worst' solution to choosing the elements of societal culture that ought to inform a science is to have them chosen by the scientists. But this clearly does not apply to the latest developments in artificial intelligence, a science which is inescapably invested with the interests of venture capital – and it doesn't apply to economics.

Instead, the values of economics filter into the society via the ubiquitous route, making, most notably in the UK and United States, taxation and government interventions anathema, this becoming more and more of the taken-for-granted reality – more and more part of the ubiquitous expertise of society in these countries – their organic face. And citizens, when facing a choice between integrity and self-interest, find it harder and harder to locate the firm moral ground. It is the free market to which 'there is no alternative' (TINA) that, to many, is coming more and more to resemble the bottom turtle. In these ways, economics has turned a supposedly value-free science into an important contributor to ubiquitous values and they are the values that are supporting the erosion of truth in society.[40]

The problem of the demarcation of science can be resolved, but it takes detailed analysis of the institution! We have economics as an example of how to pursue that analysis, but we have had to rely strongly on sources beyond our own knowledge and experience. We have suggested that economics would have a better chance of becoming a fulcrum science if it took into account the wider culture in which economic activity took place: resisting the temptation to fall back on homo economicus and the associated mathematical models, economics needs to merge with sociology. But let us be clear that we don't think, say, sociology comes ready packaged as a fulcrum science either – far from it. We see most contemporary sociology as being driven more by specific political concerns than the obsession for truth; it is just that economics has been hugely more successful in putting its political programme into practice under the guise of science than sociology. That said, economics and sociology are similar in that both have the potential to be fulcrum sciences – remember the key is obsession with truth, not with formal methods. Taking the point made in Table 1.2 about our relative level of control over our internal states, we sincerely believe that everything we have written in this book is guided by the obsession for truth. We may have failed to attain truth, but we can come close to guaranteeing that we have done our best to aspire to it.

7

Is there a chance of changing the culture?

Introduction

This book is about the threat to democracy that has been precipitated by the rise of populist politics, particularly in the United States and also in the United Kingdom. It is about the interaction of national culture and the many institutions that influence society as a whole, with science being the crucial sub-culture in respect of our argument. We have argued that the moral foundation of science resolves the epistemological problem posed by post-modernism – where is truth to be found? The best bet is that truth is to be found in the institution of science. But, recognizing that not all science fits the ideal, we have taken on the problem of demarcation and talked of fulcrum sciences and their characteristics. The detection of gravitational waves is a ready example of a fulcrum science; we have taken economics, at least, large parts of it, and notably that part that is found in the public domain, as an example of a science that is not a fulcrum science.

Economics does illustrate the way society can influence a science and the way a science can influence society as a whole, but in Western countries this influence has turned out to support the erosion of truth rather than to oppose it. In the 'post-truth era' cultural change has been heavily affected by neoliberalism and the monster of unconstrained, self-interested choice that feeds it. In the sociological model of humanity that we oppose to this, an individual is not an aggregate of selfish acts, but mere animal substance until socialized into society and then into a set of subgroups or sub-societies each providing a specific set of abilities and a characteristic set of expectations.

That politics is downstream of culture is accepted by those from both the left and right of the political spectrum. The kind of economics that has been absorbed into the culture of Western countries teaches the public, in Oreskes' and Conways' words, 'to loathe government and love the free market'. In contrast, if we want democracy to survive, we need to find ways to encourage citizens 'to loathe lies and love the truth'. Telling the truth must become more attractive to politicians from both left and right. We have to start by solving the epistemological problem of where truth comes from, and that should have been accomplished in the earlier chapters, at least to the point where our solution is a serious candidate for debate in the academic forum. But then the discussion has to find its way into the wider culture.

As explained in the Preface, this book is narrowly focused on events in the West, and the interplay of national culture and politics. The thing we are engaged in trying to understand is illustrated by some of the very recent events which went against the general drift of societal change. These events reinforce the claim that politics is downstream of culture, but these cases show culture affecting politics in a positive way, narrowly averting political disaster in the United Kingdom and the United States (at least for a time). It may be that they show there is hope for the traces of culture left over from earlier eras to contribute to the new cultural environment we want to create. If this is accepted, we can get on to the harder problem of how to manage it into the future

Traces of earlier culture and the law of conservation of democracy

Truss's fifty days and Johnson's resignation

As luck would have it, while early drafts of this book were being written, its UK-based authors were presented with interesting examples of some of the themes by the resignations of Conservative prime ministers of the UK, Boris Johnson and Liz Truss. Johnson was prime minister from 24 July 2019 to 6 September 2022 and following his resignation, Truss was elected to the prime minister's office but gave notice of her resignation just forty-five days later and left office five days after that on 25 October 2022. She is by some way the shortest-serving UK prime minister ever. Truss's fifty days were revealing in terms of the relationship between government, economics and the values of the wider population and Johnson's resignation can be seen as offering a similar moral. We'll concentrate on Truss and return to Johnson.

Conservative Party leaders are chosen by a two-stage procedure: first there is a series of elimination rounds, where Conservative members of parliament (MPs) vote for their preferred candidate, with the least favoured being eliminated in turn until there are only two remaining in the competition. There were five such rounds following Johnson's resignation, with the last two candidates being Rishi Sunak and Liz Truss. Rishi Sunak, Johnson's Chancellor of the Exchequer who had steered economic policy through the pandemic until he resigned as part of the revolt against Johnson, had strong support amongst MPs and came first in each of the elimination rounds. In contrast, Liz Truss, a cabinet minister and defender of Johnson who had refused to take part in the revolt against him, never topped the MPs' ballot and reached second place only in the final round.[1]

Conservative Party rules mean that the choice between the final two candidates is not made by elected MPs, but by the registered members of the Conservative Party, of whom there were, at the time, just over 172,000. From the start of the final round of campaigning directed at these activists, Truss was easily ahead of Sunak. She knew the activists would favour her many references back to the days of Margaret Thatcher and love her promise to carry out a radical tax-cutting programme. In contrast, Sunak's campaign took a more orthodox position on the public finances, stressing the importance of fiscal discipline and limiting inflation. Truss won the nomination with 57.4 per cent of the votes cast.

Given that the voters in this final round were unelected, and that their choice was clearly opposed to that of the elected Conservative MPs, this was a coup d'état by Conservative activists in all but name.[2] And it was a coup d'état for market fundamentalism. Truss's policies were forged in the, notorious, 55 Tufton Street, Westminster, home of a number of right-wing think tanks supported by unknown sources. They included the 'Institute for Economic Affairs' set up on the advice of Friedrich Hayek (see Chapter 6). The various representatives and members of these think tanks met frequently with Truss, her soon-to-be chancellor, Kwasi Kwarteng, and many more of her subsequent appointees, as they polished their free market, low-tax, small government agenda.[3]

Only fifty days later, after another election by MPs and with no involvement by the activists, in a turn of events reminiscent of Greek tragedy, the defeated Sunak would win the competition to be her successor, with his rivals, including Johnson, being persuaded to pull out of the race before it reached the final stage of voting which would bring back the activists to make the final decision and probably reassert their coup d'état. This does not mean that Sunak was not a neoliberal – the ideology being firmly entrenched in the Conservative Party as indicated by the 'TINA' acronym – but there are more-or-less extreme version of it and Truss was at an extreme.

As intimated, despite the nickname, there is an alternative! Look at a graph of countries' GDP per head against total taxes as a proportion of GDP and it is easy to see that there is no relationship between the two.[4] It will also be found that many of the countries in Europe where average GDP per citizen is high, including the Scandinavian counties, which are regularly reported to be among the countries with the happiest citizens, have what would count for a neoliberal as a high tax regime – around 40 per cent as a proportion of GDP compared to around 30 per cent for the United Kingdom and 25 per cent for the United States.[5] The way of being in the world of many in the Conservative Party is not based on truth, but it is based on ubiquitous economics. For many in the Conservative Party, there remains the feeling that Thatcher's low-tax aspirations have yet to be fully implemented. In other words, and despite their many years in office – at the time Liz Truss became prime minister, the post had been held by the Conservative Party for 30 of the 43 years since Margaret Thatcher's election victory in 1979 – many Conservative Party activists continue to believe that both public spending and taxes are too high; Truss knew this and exploited it ruthlessly in her bid to be the leader.

Having won the election to be the leader, Truss did not do what many were expecting and water down the tax-giveaway promises. Instead, she appointed a chancellor of the exchequer, Kwasi Kwarteng, who shared the ideas and the ideological hotbed at 55 Tufton Street, and between them they put together and presented a 'minibudget' of radical tax cuts that was designed to shatter what they saw as an economic orthodoxy that had failed to deliver the benefits of free-market economics.[6] The editor of the *Sunday Telegraph*, a right-leaning newspaper, wrote, citing academic economists to his lay readers:

> This was the best Budget I have ever heard a British Chancellor deliver, by a massive margin. The tax cuts were so huge and bold, the language so extraordinary, that at times, listening to Kwasi Kwarteng, I had to pinch myself to make sure I wasn't dreaming, that I hadn't been transported to a distant land that actually believed in the economics of Milton Friedman and FA Hayek.[7]

Truss and Kwarteng promised that their budget would trigger a revival of economic growth to 2.5 per cent per annum, sufficient to pay for the cuts and more: their policies, they promised, would transform the UK from growth-stultifying economic caution, and release the country's spirit of enterprise in the face of what they called 'the anti-growth coalition'.[8] They refused to publish the report on the economic viability of the plan that had been produced by the official body known as the Office for Budget Responsibility (OBR). Such a report is normally expected to accompany a UK budget, but they referred to the minibudget as a 'fiscal event' not a budget and claimed that this meant that no OBR report was required (rather as Vladimir Putin was, at the same time, declaring his invasion of Ukraine to be a 'special military operation', not a 'war').[9]

The events leading to Truss's resignation are well-known but let us examine some of them from the perspective of specialist and ubiquitous expertises. Liz Truss, with her loyalty to Johnson, might well have been more personally appealing to the activists than Rishi Sunak; she also sported a style in some ways like that of Margaret Thatcher: blue dress, blond hair, forceful manner-of-speaking. But the decisive tactic, coming after a series of defeats to MPs in the preliminary rounds, was almost certainly the Thatcherite tax-cutting programme.[10] Conservative Party activists, it seems, had sufficient confidence in their own economic expertise to be ready to choose a candidate for prime minister who had set out a radical economic programme over a candidate with experience of economic policy, preferred by MPs, and who had been claiming during the campaign that inflation had to be tackled before significant tax cuts could be contemplated. Economic wisdom was being treated by the activists as a ubiquitous expertise, and led by Truss and Kwarteng, they were treating themselves as the wisest of the experts.

One of the most revealing moves in the Truss-Kwarteng minibudget was the extraordinary and controversial cut to the top rate of income tax from 45p in the pound to 40p. The loss to the exchequer would be small but the ideological impact was huge: at a time when ordinary people were desperately worried about rising prices in the context of the Russian invasion of Ukraine, a giveaway to the richest people in the country was planned in the hope that this would encourage them to become more enterprising. In terms of intended impact on national culture, this tax cut stands alongside Kellyanne Conway's attempt to introduce alternative facts to national discourse – Truss was going for broke in trying to redefine the level of inequality that citizens should count as normal.

Going back to the economics, the idea behind the tax cut is that a significant proportion of the money given away to the rich finds its way into enterprise-related activities, thereby boosting economic growth. In reality, because the rich already have plenty, their additional income is often saved rather than spent (unlike the poor, who will usually spend the vast majority of any additional income), and evidence to support this 'trickledown' theory is hard or impossible to find.[11] A couple of weeks into the uproar, on 20 September, even US President Biden is reported to have said on Twitter: 'I am sick and tired of trickledown economics. It has never worked.'[12] The debate over the cutting of the top rate drew attention to calculations showing that the overall package of cuts was regressive, with rich people gaining the most. *The Observer* (a left-leaning newspaper) reported: 'Households in the top 5% of wealth distribution will gain on average, more than £8,000 a year as a result of Kwarteng's changes; those in the poorest 10% get nothing' (25 September). The furore was such that even the International Monetary Fund (IMF) took the unusual step of issuing an openly critical response to the minibudget on 28 September, saying it would 'likely increase inequality'.

For Truss and Kwarteng, however, these calculations and criticism were dwarfed in importance by the reaction of another set of experts – financial investors. Pension fund managers and other professional investors essentially decided that they didn't believe that the promised growth would materialise and that, as a result, UK government debt would have to increase sharply in order to make up the shortfall created by reduced tax revenues. In practical terms, this meant that the pound collapsed and so did the market for bonds, pushing up the price that investors demanded for holding government debt, making it hugely more expensive to service. This, in turn, forced the Bank of England to intervene, buying government debt in order to restore stability, and giving rise to what became known amongst City finance traders as the 'moron premium'. This would soon lead to the reversal of the policies and the resignations of the would-be culture transformers. The champions of the market had failed to understand the most important market their budget was addressing – the investment market. It is difficult to imagine that these concerns were not raised – or at least could not have been encountered – if Truss or Kwarteng had been prepared to take the economic orthodoxy seriously. Unfortunately, their unshakeable belief in the market fundamentalism of ubiquitous economics meant that neither they nor their activist supporters had any doubts that the policies would be welcomed.

The question we asked in Chapter 6 was about economics as a science and the location of economic expertise. We have seen both Tory activists and Truss bet on their own expertise in ubiquitous economics when it came to putting radical tax cuts in place. Truss went through a month of utter humiliation but was still laying claim to her economic wisdom in her surprisingly upbeat resignation speech presented, on 25 October. Addressing the media outside 10 Downing Street she said:

> We simply cannot afford to be a low-growth country where the government takes up an increasing share of our national wealth.

She was still unashamedly equating bigger government with lower growth, on the face of it, a technical claim belonging to the science of macroeconomics which, as we have seen, shows there is no such correlation, and a classic example of the misrepresentations found within ubiquitous economics.[13] Likewise, the editor of the *Sunday Telegraph* was betting on his expertise and writing in his newspaper for his readership in the confident tone of expert-to-expert that one might find in small groups of physicists – 'We *Sunday Telegraph* readers all know that Hayek and Friedman were right and at last there are some politicians who have grasped it too.' Actually, of course, all of us are using economics as though we own it, in a way we would never use, say, physics; all of us have been sucked into the not-quite-science of ubiquitous economics. We are acting as though taking part in a democratizer's dream where esoteric expertises have become ubiquitous expertises.

Truss's confession

In a remarkable 4,000-word article published in the *Sunday Telegraph* on 5 February 2023, Truss tried to explain why she failed so spectacularly in her attempt to push British politics so much further to the right. Incidentally, she seems to have had no sense that she was the beneficiary of an effective coup d'état based on the votes of less than 50 per cent of Conservative Party activists, instead describing her ascent as a matter of political acclaim:

> My party agreed that we needed to stop drifting in the direction of a higher-tax, higher-spend economy being choked by ever more regulation, which was causing sluggish growth, low productivity and a dampening of enterprise and innovation. My plan to get Britain back on the right trajectory was popular.
>
> So it was that I won the leadership election with a clear mandate from my party in the country and, by the close of the ballot, the backing of the majority of MPs declaring a preference. I entered 10 Downing Street determined to deliver the bold action I had promised, with the economy and energy as my key priorities at the top of the in-tray.

Truss recognized that her crashing of the economy was important but claimed – oddly for a prime minister hand in glove with her chancellor of the exchequer – that she had not been informed of the dangers to pension funds:

> Brewing in the background there was an issue relating to pension funds, which neither of us had been made aware of – a problem that would ultimately bring my premiership to an abrupt and premature end because of the panic it induced.
>
> At no point during any of the preparations for the minibudget had any concerns about liability-driven investments (LDIs) and the risk they posed to bond markets been mentioned at all to me, the chancellor or any of our teams by officials at the Treasury. But then, late on the Sunday night, came the jitters from the Asian markets as they opened. I was alerted to this on the Monday morning, at which point the Bank of England governor was wanting to make a statement on LDIs.
>
> Readers will not be surprised that, given their impact on events, since leaving office I have spent some time looking into LDIs. I was shocked by what I discovered.

This extraordinary account of how she came to crash the economy reinforces the claim made earlier that economic debate would be stronger if it relied less on assertions based on the theorized interactions of self-interested individual acts and took more account of mainstream econometrics and other social science techniques such as market research and participant observation. Before proposing the reforms, Truss should have been willing to listen to those economists in the OBR, and elsewhere, who could have led her to understand the impact of the minibudget on pension funds and their consequences for investment decisions; instead, she claimed to have found out about this only afterwards. The market may be impossible to predict with complete accuracy, but had advice been sought, it would surely have said that unfunded borrowing on that scale envisaged could only have a negative effect. In other words, even if we can't predict the future exactly, it is possible to foresee a window of more or less likely possibilities, especially if we allow ourselves to look into what market-makers are thinking – a quick probe (see Chapter 6) would have done it. As she admits:

> Knowing what I know now, undoubtedly, I would have handled things differently. I underestimated the extent to which the market was on edge and, like many others, was not aware of how fragile our system had become.

But Truss is also aware that politics is downstream of culture, and she regrets the lack public education in what we have called ubiquitous economics:

> Large parts of the media and the wider public sphere had become unfamiliar with key arguments about tax and economic policy and over time sentiment had shifted leftwards. This is partly because we Conservatives had failed to make these arguments enough since 2010 …
>
> This high-spend, high-tax policy view does not just prevail in the UK. We were also swimming against the international tide. There was a concerted effort by international actors to challenge our Plan for Growth. The IMF commented on distributional aspects rather than market stability which it is hard to conclude was anything but politically motivated.
>
> Then there was the intervention from President Biden, who publicly voiced his disagreement with our economic policy, stating: 'I wasn't the only one that thought it was a mistake.'

Truss, as was already evident in her resignation speech, was unshaken in her economic theories by her failure and 100 days later was still unmoved. She complains that median incomes in the United Kingdom are well below those in the United States, Switzerland and Norway but explains this is because:

> We have not done enough over the past decade to make the arguments for a lower-tax, more deregulated economy …
>
> [We must] persuade the wider electorate that ever-higher taxes, an ever-more restrictive regulatory regime and historically high levels of state spending are the key underlying causes of poor economic performance.

But according to the OECD, the estimate for percentage of GDP going into taxes in 2021 in the four countries she mentions are: United Kingdom, 33.5; United States, 26.6; Switzerland, 28.0; and Norway, 42.2.[14] As mentioned above (p150) there seems very little evidence that lowering taxes in wealthy countries increases wealth – some of the richest have low taxes and some have high taxes. Why is the low-tax, polysemic, *slogan*, not being *discharged* in the direction of clarity of meaning, by, first, saying this is a theory not well-backed up by evidence, and second, saying in the meantime low taxes mean lower wages and fewer resources in the public sector, at least in the short and medium terms, so prepare for greater strains and more strikes in the health service and the education sector where these are funded by taxpayers, and prepare for a long-term future where personal income is diverted to insurance schemes for health and for private education for those who can afford it leaving those who can't with ever poorer services. Perhaps that is what the electorate would want, perhaps not, but that is how the low-tax slogan would be presented in a veritocracy leaving voters knowing what the future most likely holds in consequence of their choices.

UK national culture

Much of the story of Truss's fifty days really belongs in Chapter 6 on economics, but we have located it here because there is another moral that carries some hope about democracy in the UK which applies to both Truss's and Johnson's resignations. The proximate cause of Liz Truss's resignation appears to be the reaction of the markets to an uncosted tax-cutting plan but there was another cause – the same cause that brought down Prime Minister Boris Johnson before her. Opinion polls had shown that Johnson would lose a general election, and lose heavily, as the British public would not be prepared to vote for a prime minister who was widely seen to have been telling a series of lies. Even if his own party would have been prepared to pay the moral price of his charismatic presence, the public at large would not. Johnson had surpassed even the high threshold of perceived lying for a charismatic leader.[15] In the Truss case, opinion polls reacted strongly to the cut in the top rate of income tax and the evidence that the entire package favoured the rich over the poor. It became rapidly evident that Truss's package, attractive as it was to Tory activists, would result in a huge loss of seats to Conservative MPs and a landslide loss to Labour should there be a general election in the near future.

Now we have to make an informed guess. Huge and rapid swings of popular opinion of this sort are not likely to have been caused by economic calculations but by moral judgement. Johnson's loss of electoral viability came about because of his display of what was seen as moral irresponsibility – the perception that he was lying. In the case of Truss, she was promising trickledown in the longer term, but her immediate actions pointed to a lack of moral judgement in respect of what was currently acceptable in the way of inequality. The cut to the top rate of income tax was unlikely to have much economic effect but it had a huge symbolic impact: it was saying that the UK was ready to embrace more and more inequality for the sake of growth. The electorate, it seems, along with the IMF of all people, were agreeing that this level of inequality, and, even more important, inequality-flaunting, was not acceptable in absolute terms irrespective of any economic calculation. The population, of a country with a very thin record of political violence, was ready to stage a revolt at the ballot box.

This interpretation is supported by the Truss government's own actions. Truss began a programme of reversal of the tax-cutting measures presented in the 'fiscal event', but the first of these reversals concerned one of the least fiscally impactful – the top rate cut. This reversal was unlikely to have much impact on the markets (which it didn't). Here was direct democracy in action: opinion polls showed that the British public found the minibudget so unacceptable that it would lead to huge electoral losses and the first response of the government was to reverse the most *morally* unacceptable measure, only later being forced into U-turns on more significant fiscal measures. So, in spite of what Truss claimed, this was not an ill-educated public failing to understand the science of economics, but it was a public exercising its moral right to refuse more inequality – the downstream consequence of the proposed policy changes. What was being demonstrated in both the Johnson and the Truss resignations was that, even in the short term, the politics of the time really was downstream of culture. In the case of Johnson, he still could not be seen to lie beyond a certain level of shamelessness without suffering political rejection, even though the level of politically acceptable untruth was much higher than it had been only a few decades earlier, and even though the level was rising fast: Johnson had exceeded the 'speed limit' for being perceived as telling lies in public places. As for Truss, even though the level of inequality in Western societies had grown to what, in the decade after the Second World War, would have been unimaginable, and even though it was still growing fast, she had exceeded the speed limit for inequality flaunting in the UK. Truss had demonstrated how ubiquitous economics could draw praise from sectors of society such as the editor of *The Sunday Telegraph*, but had also revealed that society as a whole had yet to reach that level of acceptance of market forces unconstrained by moral forces.

Truss, herself seemed unable to grasp this. In her 'confessional' article in *The Sunday Telegraph* of 5 February she wrote:

> I underestimated the resistance inside the Conservative parliamentary party to move to a lower-tax, less-regulated economy. The furore over the reduction of the top rate of income tax was testament to this. In the overall package of measures, it was – in fiscal terms – little more than a rounding error, equivalent to 0.2 per cent of government spending. Even though the measure was economically

> sound, I underestimated the political backlash I would face, which focused almost entirely on the 'optics'.

For Truss, a public rejection of a further gratuitous increase in the gap between rich and poor was a matter of 'optics' – the way things look. A strangely dismissive way to talk for a party that had based its campaigns on slogans, but also a very impoverished sense of morality, in which appearance is all. If, however, we see what happened to Johnson and Truss as reflecting some deeper civic morality then this holds out some hope for the public's potential to understand the benefits of veritocracy.

The US 2022 midterms

The 2022 midterms tell what is in some ways a similar story. Democrat incumbent Joe Biden told citizens that democracy was at stake in the midterms because many Republican candidates were explicitly endorsing Trump's false claim that he had won the 2020 election while it had been stolen by the Democrats – the 'stop the steal' slogan. As it turned out, at the time, just enough Americans seem to have understood Biden's account of democracy to have defeated quite a few of the Trump-inspired candidates. The Democrats held on to the Senate even though the historical precedent in midterms is for the incumbent party to lose heavily and even though Biden's policies were not working well, and Biden himself was an unpopular candidate according to the polls. Enough of the US population had, it seems, understood democracy well enough, and wanted it badly enough, to save it, at least for the moment.

But why was it such a thin majority of the population that understood and wanted democracy? On 15 November, the day it was reported in the UK that Kari Lake, the Republican candidate for the governorship of Arizona and a prominent supporter of the 'stop the steal' campaign, lost to a Democrat, the BBC reported that at least 125 election deniers had won races for the House, Senate and state governorships. We are arguing, of course, that national culture, which encourages the idea of individual freedom expressed in market fundamentalism was a major contributor. Absorb this idea and the 'moral values' of the market replace wider obligations. As individual freedom becomes the only value, so the general population are free to choose lies over truth if that serves their interests better. As Deirdre McCloskey suggested (*Harper's Magazine* reported above p122), the business schools went from trying to make managers into professionals – like lawyers, professors and doctors, who acknowledge a deep ethical commitment to their clients, their students, their patients – into making people with a kind of 'screw you' attitude. (It is easy to see what this means for the analysis put forward in Figure 3.2; the downward arrows arising out of national culture all stress self-interest and market forces and the upward arrows no longer point to truth and other moral values.) It is to the erosion of truth within the national culture that we have to look to explain the huge numbers who voted for Trump candidates in the 2022 midterms, even though the surviving moral and political understanding of the slightly larger majority of American citizens saved Congress from an overwhelming Trump-supporting majority (but now we know this reassurance lasted only another two years).

Of course, the culture we are looking at is bolstered by the celebration of the rugged individual surviving on the frontier as in *The Little House on the Prairie*, Western movies and their successors, the well-developed scepticism about government that can be found in American political philosophy, and since the 1970s, the surprising public acceptability of growing inequality in Western democracies, which seem to have forgotten the lesson of earlier revolutions: that the sufficiently downtrodden will endorse any doctrine that might make their lives better because they have nothing to lose. It is hard to understand that since the decade of the 1970s, poor Americans have been getting poorer not richer while rich Americans have been getting richer and yet so few people have understood what this means for our political culture. The threat of this culture to the survival of truth is what we have been trying to show. But the incidents we have described that took place in 2020 suggested (for a time) that all was not yet lost: it appeared that *homo economicus* did not describe society in the absence of *homo moralis* and economic models based on *homo economicus* are not models of society.[16] Johnson fell because the public rejected what they saw as his lies, as revealed in opinion polls, interpreted by his own MPs, and the same was true of Truss's proposed injection of more inegalitarianism into the national culture. But this triumph of *homo moralis* no longer seems to apply to the United States.

The meaning of the Johnson, Truss and (maybe) Trump failures

If Truss and Kwarteng had succeeded, they would have injected into British culture a readiness to accept a further ratcheting up of inequality for the sake of a (dubious) promise of long-term economic growth, just as Johnson's continuation in office would have meant the further establishment of an appearance of lying as an acceptable and regular feature of public office in the UK. If Trump's candidates had overwhelmingly succeeded in the 2022 midterms, it would have been another triumph for the erosion of truth.[17] But the reaction of the British population to Johnson and Truss indicates that a considerable proportion of UK citizens do still understand moral culture and are not ready to sacrifice it to political slogans or economic technocracy beyond a certain point.[18]

The landslide victory of the UK Labour Party in the July 2024 election, said to be the biggest reverse ever in electoral fortunes between Labour and Conservative in the UK, and leaving the Conservative Party with their smallest number of MPs ever, showed that this change of heart was not a flash in the pan: the citizenry voted for a leader who presented a calm and quiet 'efficient' demeanour, lacking in charisma, while the Labour Party had no striking policies on offer. The UK electorate recognized the moral and technical chaos that resulted from their earlier choice of Johnson (which in turn led to Truss) and demonstrated with their votes that they did not want more of it. This is the positive operation of the 'law of conservation of democracy', which, to repeat, follows the form of the first law of thermodynamics, in holding that no more democracy can be taken out of a society than is put into it.[19] In simple language, citizens don't know how to act nor who to vote for, when democracy is challenged by political leaders unless they have already been taught to understand and appreciate the meaning of the culture of democratic freedom. The reaction of UK citizens, as gauged by the opinion polls and followed up by the actual election landslide, supports the idea that the culture of democracy is still alive in the UK.[20]

Of course, the UK citizenry had earlier voted *Johnson* in with a landslide victory, so it shows that the UK electorate is vulnerable to showmanship over truth and technical efficiency and more will be needed if this change of heart is to be enduring and not vulnerable to showmanship on future occasions.

Worryingly, the results of the 2022 midterms in the United States were more equivocal: democracy survived, but only just. It turns out that there are still a large number of Americans who adhere to the idea that the 2020 election was stolen, not least Donald Trump himself. Trump, at the time of writing (September 2024) is standing for president once more and 'the steal' is just as much part of his platform as it ever was. Furthermore, he has said he will not accept the result of the next election if it does not go his way. If it does go his way, there are rumours that he will pardon the Capitol rioters. The more people say the election was stolen, the more does the credibility of that alternative fact grow, accelerating the erosion of truth in general. Furthermore, 'Project 2025', a big document produced by the right-wing think tank The Heritage Foundation, including many Trump followers among its authors, and which claims to be a blueprint for the next Republican government, prescribes, among other things, the replacement of many trained civil servants with political appointees. At the time of writing, less than two months before the American election, half of the American population are still saying they will vote for Trump: the chaos he caused and his elementary school playground level of argument – for example, immigrants eat peoples' pets and that he always tells the truth whether it gains him votes or loses them – does not seem to bother them. This difference with the UK can only be explained in terms of political culture. (We leave this paragraph in its pre-election form so as to dramatize the change that has unfolded since it was written.)

This analysis has an uncomfortable implication for the UK 'remainers' among us, including the authors, which the obsession with truth obliges us to admit. Throughout this treatment we have taken the Brexit decision in the UK to have arisen out of lies about its economic implications and slogans such as 'take back control', which were in turn based on incorrect claims about how much control the UK already had within Europe, how much it would regain once it left and whether immigration was good for the economy or simply priced British workers out of jobs. This is the antithesis to the kind of policymaking we would want to see in a veritocracy – which would shift from slogans to their cashed-out consequences. But the above analysis of utility also allows that, for a good part of the population, the preservation of national identity could be seen as an absolute good. Perhaps this part of the population *was* ready to sacrifice any potential national economic benefits for the sake of their idea of Englishness, or Welshness. Perhaps, irrespective of their economic impact, the very presence of immigrants, their languages and their foods, was an absolute bad for those who voted for Brexit, when set against the 'moral' notion of Britain as a unique cultural entity. The argument seems to allow for some or all of those who have suffered economically with Brexit to feel, nevertheless, that their lives are better now that the perceived *cultural* danger of integration with Europe has been reduced. It seems to the two British authors of this book, noting the popularity in the UK of Italian, Chinese and Indian restaurants, the number of MPs, including the previous prime minister and the current leader of the opposition, celebrities and sports stars from ethnic minority backgrounds, and the gladly welcomed proportion of overseas contributors to the medical, caring and other professions, that this probably is not true, but we cannot ignore the possibility.[21] If it is true then this is another example of economics being downstream of culture, in this case nationalism as a feature of culture, with utility maximization having to take the preservation of ideas about national cultures into the calculation, not just the ever more obvious economic losses that come with economic isolation.

How to change society's culture

At the end of the introductory chapter we suggested that there were three ways to explain the erosion of democracy: the institutional, the material and the cultural. Here we have concentrated on the cultural but many changes have contributed to the creation of an environment in which the erosion of truth could appeal to the electorates of democratic societies. The growing gap between rich and poor is probably the most important.[22] We authors have no means of changing these things except by changing the culture that has brought us to where we are. And if that seems an impossible task, we should remember that all cultures change over time, with change as likely to come from the accumulation of small events as from the effect of single major incident. One place these incremental changes might be made is in the early socialization and education experiences. As has been pointed out, both left and right believe politics is downstream of culture. To repeat, even Liz Truss blamed cultural and educational deficits for her demise:

> We have not done enough over the past decade to make the arguments for a lower-tax, more deregulated economy …
>
> [We must] persuade the wider electorate that ever-higher taxes, an ever-more restrictive regulatory regime and historically high levels of state spending are the key underlying causes of poor economic performance.

We agree that education is the key. There is nothing new about this nor about our concern about civic education's demise. According to Yascha Mounk, writing in *The Observer* (Mounk 2018), the early US presidents were well aware that self-government required a knowledgeable population. Thus:

> For the first centuries of the republic's existence, this emphasis on civic education shaped the country. Parents sought to raise tomorrow's citizens, competing with

> each other to see whose four-year-old could name more presidents … Civic education in all its forms stood at the core of the American project – as it also did in, say, Britain, Germany and Scandinavia. Then, amid an era of unprecedented peace and prosperity, the idea that support for self-government had to be won anew with every passing generation started to fade. Today, it is all but extinct.

Philip Mirowski (forthcoming) argues that the popularity of neoliberalism in the United States has led to the government abandoning its responsibility for civic education, the only ideologically acceptable solution being to leave such things to the market. But this has to change unless Western democracies want to see the internal contradictions of a democracy based on absolute freedom to choose leading to its imminent decline. If democracy is to survive the idea that free markets can give rise to a moral order has to be seen to be crazy, as Hayek saw. However ideologically committed we are to freedom, the last decade in the West has provided undeniable lessons. And now we need a 'nanny state', ready to revive civic education, if we want what we call 'civilization' to endure.

There is empirical evidence that civic education can work in promoting cultural change in the narrower domain of relationship and sexuality education. School-delivered programmes have been effective in reducing the proportion of young people engaging in what are classed as 'risky' sexual behaviours. The outcomes have not been uniform but the evidence that civic education can have an impact is positive.[23] We have to gamble that it can work in respect of more general understandings of how society works so we do not fall foul of the law of conservation of democracy.

Remember the urgency of the problem! The V-Dem project at the University of Gothenburg has been charting the changing nature of political systems around the world since at least 2017, with annual reports tracking how countries either democratize or 'autocratize'. The 2023 report makes grim reading, with the executive summary offering the following key messages:

- Advances in global levels of democracy made over the last 35 years have been wiped out.
- The level of democracy enjoyed by the average global citizen in 2022 is down to 1986 levels.
- The world has more closed autocracies than liberal democracies – for the first time in more than two decades.
- Freedom of Expression is deteriorating in 35 countries in 2022 – ten years ago it was only 7.

- Government censorship of the media is worsening in 47 countries.
- Government repression of civil society organizations is worsening in 37 countries.
- Quality of elections is worsening in 30 countries.

(Papada et al. 2023: 6)

Of particular concern is the fact that the United States is included, alongside Brazil, Ghana, Greece and Poland, as democracies that are 'engulfed in autocratization' (pp19–20), with the analysis further noting that this weakening of democracy has, in some cases, led to complete breakdown and the reclassification of countries such as El Salvador, Hungary and India as electoral autocracies. Once more, we leave the earlier draft sentiment in place so as to emphasize the most recent developments.

But what chance do our arguments have in persuading people that truth is important? As we said at the beginning of Chapter 1, the first problem we have to solve is a philosophical one: where to find truth. We have to get ourselves out of the post-modernist pit. As Wave-Two analysts of science, digging the pit has been, to some extent, our responsibility. But, unlike the science warriors, we don't want to simply fill it in again by going back to Wave One. That would be a defeat for fifty years of analysis of science and the related field studies. It would also be a failure of veritocracy: to go back to Wave One would be to fail to tell the truth about what has been discovered. And the whole of this book is about trying to tell the truth! But we think we can get out of the pit with Wave Three. Table 7.1 summarizes our ambitions.

Table 7.1 summarizes the audiences we hope to reach, either directly, or, if we are to have a real impact, indirectly, via our direct audiences' good offices. The first column of the table reminds us of the attempt we have made to solve the epistemological problem. The second column reminds us that we are duty bound to try to persuade our academic colleagues to think about it seriously. The third and fourth columns point out that we need others to bring the approach to the wider society. The fifth column describes what would happen if all this came to pass. We'll look at the columns starting from the left:

Table 7.1 Cultural change as we hope it might arise out of our analysis.

Solve the epistemological problem: (Get out of the pit)	**Convince sociologists philosophers and perhaps political scientists**	**Strike a chord with powerful actors?**	**leading to changed culture in the home and teaching democracy in schools**	**which will change the cultural environment of politics**
	This book Articles Conference presentations	Journalists Commentators Broadcast media Democratically inclined business enterprises Political parties Educators	Vaccination Climate change as object lessons that could appeal to a wide audience along with a new citizenship syllabus Teachers	Politics is downstream of culture **In the new culture truth-telling and integrity in general become integrated in society's ubiquitous expertises and politically attractive to the political classes and to voters, while polysemous slogans lose their force** Democracy survives the law of conservation

Columns 1 and 2: Persuade immediate academic colleagues

Our way of getting out of the pit is represented in the first column of Table 7.1, which draws again on the 'turtles all the way down' story. Wave Three locates the turtle that supports the academic and scientific world, and hence the political world, in the domain of the moral. It is a matter of evaluating and ordering society's institutions. We think that the institution of science is where truth is most securely located, not because science is guaranteed to discover the truth, but because the institution of science – science proper – is defined by its aspiration to tell the truth. That, we say, is how we establish a foundation for the moral values needed to support societal and political culture in which the truth matters. Truth is the foundation of veritocracy. And the institution of science is the best object lesson for a veritocracy.

The second column describes our professional duty as social scientists employed in a university. Our duty is to try to convince our disciplinary colleagues that there is a solution to the epistemological problem and that our way of getting out of 'the pit' is a good one. The book is meant to convince those of our colleagues whose work touches on the sociology, philosophy and history of science, but this will be no easy task. Even amongst those who have recognized the dangers of the pit, a significant proportion, particularly those in the philosophy of science, think that the only way out is Wave One, where science was once thought to live a semi-autonomous existence divorced from the doubts and cultural forces which appear when science is examined more closely, as it was in Wave Two.

On the other hand, many of our more immediate colleagues exhibit the opposite problem: they celebrate Wave Two and either do not feel the dangers or do not see how they are contributing to them by focusing only on the imperfections of science. They do not seem to experience the new political storms as related to their professional life and so do not feel the need to reflect seriously on how their work feeds into them; at best, they concentrate on the causes that don't relate to their work on science. They are not interested in finding new ways to analyse the positive features of science or in identifying the distinctive features that differentiate science from other activities. Among other things, we wish we could persuade those colleagues of ours who, in the late 1990s, enthusiastically endorsed the parents' revolt against the MMR vaccine, to admit, in public, that were wrong. Ironically, they have in their possession a very important key to the prison of post-modernism and neoliberalism in virtue of their position in the MMR and similar debates. Admitting they were wrong could turn that key.[24]

Whatever, we must keep trying to use the methods through which science more regularly advances. Since the 2002 paper that launched the 'studies of expertise and experience' (SEE) approach, we have completed research projects, presented at conferences and published peer-reviewed journal articles and books. Some of these publications have been widely cited, suggesting a degree of impact but our sense is that this is mostly outside the core Science, Technology and Society (STS) community. Indeed, if anything, we have been surprised by the bitterness of the reaction within STS to the idea that science has positive value.

The popularity within the STS community of the democratization of science idea, irrespective of its alignment with populism, is worrisome as it suggests an unwillingness to consider alternatives. Thus, we continue in our attempt to convince this community to modify the monotonic programme of critique and to work out what is good about science. After all, reflection will surely lead to the recognition that nearly all of us would rather have a world with science than a world without it, or a world in which scientists' findings have to adhere to the whims of a dictator (or a political majority).

The problem with the critique of science that leads to the 'democratization' of science and the confirmation of epistemological rights to 'lay experts' is that it is a populist programme in the absence of any worked-out limits on the role of citizens (the problem of extension). There are no 'lay experts' when it comes to the technical content of science – science is essentially an elite activity (like the law). This is not to say that there are not experience-based experts in certain domains who can provide valuable input to scientific research, but they should be seen as part of the technical elite with their input evaluated for each specific case, as with all contributors to a technical domain.[25] The STS community must solve 'the problem of extension': in so far as those outside the scientific community have rights to contribute to the content of science, how far do those rights extend and where are the boundaries drawn? If these questions are not answered, then science collapses into politics. We have tried to reconcile these matters in Chapters 4 and 5, and in Chapter 6 gave an example of how not to reconcile them.

Column 3: The powerful actors

The next influential groups, the journalists and political commentators mentioned at the top of the third column, are crucial for all the reasons discussed in Chapter 6: political change of the kind we want to bring about currently comes not so much from working directly with political parties but by changing the culture within which they operate. Here we could learn from Anthony Fisher's interaction with Friedrich Hayek but unfortunately there are few very rich people likely to want to establish truth-championing think tanks. But serious journalism still has an important role to play in the West. If the conversation about the value of truth and the role of science in it is to go anywhere, that institution will be a key actor.

Some idea of what is needed is provided by the very success of ubiquitous economics; it reveals the scale of the challenge, both in terms of the diversity and volume of interventions needed and the time period over which such an effort must be maintained. This is clearly not the kind of task an individual, or even small group of individuals, can sustain. If it is to succeed, then it will be via building networks and by others seeing connections with their own concerns and interests. Awareness is the first step on such a journey and academic journals are only read, if at all, by academic colleagues. Crucial are boundary organizations that bridge the gap between the university, policy and the media. As with all science, there is an obvious ‘Mathew effect’ in these areas, with scientists who move in these circles, either by dint of their own efforts or by virtue of being located in well-connected institutions, being the familiar names and faces that policymakers and journalists turn to when comment or expertise is needed.

On the other hand, there is already a growing literature that addresses concerns not dissimilar to those in this book and to which the argument for veritocracy is at least partially aligned (even though the idea that science is the key to political salvation is pretty-well unique to us). The rise in publications about, and preoccupations with, populism and its consequences, no doubt reflects the rise of populist leaders in countries around the world and the attention this has been getting within political debate and political theory. There has been a rapid increase in the books that deal with these topics and are also aimed at something wider than a narrow disciplinary audience. At the end of Chapter 1 we mentioned a list of fourteen that we would present in this chapter; they are listed in Table 7.2.

With many variations on the same theme, the books argue that democracies around the world, but particularly in the so-called advanced economies of the United States and Europe, have, as a result of demographic and other changes, left their citizens less committed to the ideals of liberal democracy than they once were and more willing to support radical candidates. Many of these books have been reviewed in the broadsheet press, suggesting some penetration into the wider culture. In the UK, this suggestion is reinforced by the topic chosen in 2023 for the annual, UK, Reith Lecture Series: ‘Our Democratic Future’.[26] Taken together, these suggest that interest in challenging politicians and informing public debate exists.

In the case of economics, it is plausible to imagine that, following the horrors of the Great Depression and the apparent success of Keynesian economics,

Table 7.2 Authors who fear the death of democracy.

Jan-Werner Müller (2017) *What Is Populism?*
Yochai Benkler, Robert Faris, and Hal Roberts (2018) *Network Propaganda: Manipulation, Disinformation, and Radicalization in American Politics*
Roger Eatwell and Matthew J. Goodwin (2018) *National Populism: The Revolt against Liberal Democracy*
Steven Levitsky and Daniel Ziblatt (2018) *How Democracies Die*
Wendy Brown (2019) *In the Ruins of Neoliberalism: The Rise of Antidemocratic Politics in the West*
Yascha Mounk (2018) *The People vs. Democracy: Why Our Freedom Is in Danger and How to Save It*
Paul Collier (2019) *The Future of Capitalism: Facing the New Anxieties*
Timothy Snyder (2019) *The Road to Unfreedom: Russia, Europe, America*
Jason Stanley (2020) *How Fascism Works: The Politics of Us and Them*
Robert Talisse (2020) *Overdoing Democracy: Why We Must Put Politics in Its Place*
Anne Applebaum (2021) *Twilight of Democracy: The Seductive Lure of Authoritarianism*
Michael J. Sandel (2021) *The Tyranny of Merit: What's Become of the Common Good?*
Steven Levitsky and Daniel Ziblatt (2023) *Tyranny of the Minority: How to Reverse an Authoritarian Turn and Forge a Democracy for All*
Peter Turchin (2023) *End Times: Elites, Counter-elites and the Path of Political Disintegration*

politicians and the public now needed to be persuaded of the threat, put to them by market fundamentalists, that any state intervention was dangerous. Today, in contrast, it is hard to imagine that either politicians or citizens committed to democracy are unaware of the dangers posed by their more populist opponents. What is still missing seems to be any shared, positive vision for what a revitalized democracy might look like and what a civic education programme designed to support it might include.[27] If anything, the mainstream parties are at risk of following the agenda of the more populist opponents in a futile attempt to meet their supporters half-way.[28] It is this absence in the current debate of a positive case for liberal democracy, as opposed to limiting the damage done by its opponents, that this book and those like it seek to fill. It is only if these ideas, or ideas like them, start to circulate more widely in the upstream culture that the values and practices of pluralist democracy can be restored.

Column 4: Socialization in school and home

The general public, we hope and believe, is still open to persuasion that science's commitment to honesty offers the best example for decision-making of all kinds and offers the best technical advice in respect of the observable world. The esoteric argument about the centrality of moral truth to science as an institution is not something that is likely to appeal to the public at this stage, but maybe more concrete issues can open the gate to the more subtle argument. Science has had a good press recently in consequence of the COVID pandemic and fears over global warming (though this is now under threat in the United States). The public when faced with reported death rates from COVID in different countries seem to have understood that the United States and the United Kingdom initially handled the pandemic badly. Ignoring medical advice for the sake of political sloganizing, including the Trumpists' refusal to wear masks at political rallies, was not the best way to run a country. Perhaps reports from Bolsonaro's similarly ineffective approach in Brazil bolstered this interpretation of what was happening the United States. It is not hard to find commentators who believe that Trump's handling of the pandemic turned the election against him in 2020. At the same time the idea that climate change caused by humans is real is also gaining traction, perhaps because of the recent increases in extreme weather events even if the statistics are only just becoming strong enough to link the two with anything approaching certainty. On the other hand, the interests of the parties rejecting climate change are so obvious that anti-global warming rhetoric, though still attractive to the far right, is not likely to be an election winner.[29] It does seem that many of the general public concluded over the past few years that diminishing the input of experts into these problems for the sake of immediate political or financial gratification does not lead to the best long-term solutions for them, their societies or for future generations. But this idea is under threat again in the United States.

This could be said to be a Wave-One argument – depending on the success of science rather than the moral worth of scientists: 'the scientists were more right about vaccines and more right about global warming'. But our model of science is also about the way to bet when it comes to substantive issues: choosing experts who aspire to the truth is more likely to get to the truth than choosing those who aspire to something else. As we put it elsewhere:

> The very notion of expertise contains the notion of utility. If no kind of expertise was ever efficacious, the meaning of the term 'expertise' would change and it

> would become something like 'opinion'. So, in spite of the major claim that the main defence of expertise has to be moral or commonsensical, we cannot avoid hoping that expertise is also efficacious and that part of the way of being of experts has to be to try to be useful; what we are saying is that they have only to try, not to succeed, to be doing good.
>
> (Collins and Evans 2017b: 25)

What we are saying is that vaccination and global warming are being accepted as places where the bet is paying off, as it must pay off a lot of the time if science is to be counted as expertise. We are saying, or hoping, that the pay-off in these two cases is gaining enough traction to open the public to idea of the value of science. But for the positive conclusion to be robust and not a perpetual hostage to fortune, the public will also have to understand that science is not the unfolding of an inexorable logic, but the best choice given the uncertainty of both knowledge and policymaking.

Moving still further up the fractal model and into the larger social institutions of school and family, we need these changes in the direction of veritocracy in popular culture to be founded and then reinforced through civic education in schools.

As with the debates mentioned in Chapter 2 about the best way to understand science, something similar needs to be said about pluralist democracies. Just as the public understanding of science would be enhanced by a more realistic understanding of what science consists of – i.e. the craftwork with integrity model – so any defence of pluralist democracy will need to acknowledge its difficulties as well as its successes. As with science, it may well be that the aspirations and values of democracy are its most important qualities, with the goal of civic education being to both explain what these are and demonstrate why, for all its problems, democracy remains the least-worst form of government. In other words, a democracy has to take responsibility in its educational institutions for teaching its citizens about what democracy is because if they don't democracy will not survive the ballot box.

We *hope* that by teaching citizens from an early age what democracy is, it will survive the ballot box because citizens will understand and recognize when it is not being practised properly. For this to be realized, it will require some awareness and uptake of the ideas within the relevant policy and professional educational circles. In addition to the standard material about respect for pluralism, the need for checks and balances, the importance of the rule of law and a free press, civic education in a veritocracy would also include a course called 'Science and Democracy for Citizens'.

This course would explain how science really works. It would no longer be good enough to teach the old myths where the 'hostage' is scientific certainty, and the 'fortune' is science practised at the speed of politics. Instead, the 'craftwork with integrity' model of science would be taught. It would be crucial to explain that scientific advice is bound to fail from time to time because of the speed of political decision-making. Nevertheless, science remains the best bet because of the commitment to trying to find the truth, which is stronger in science than in all, or almost all, other institutions. How science works to maximize the chances of the bet, with safeguards against expectancy effects and other prejudices, with internal criticism among small trusting groups, with mutual agreement in such groups about assumptions and with painstaking checking of results, would be explained with examples.

Science, like the law, has to be conducted within bounded expert groups and its elite status is just as compatible with democracy as the elite nature of supreme courts is compatible with democracy – the elite nature of these institutions is a condition of democracy, not a threat to democracy. The threat to democracy comes from linking the law too intimately with politics and the same applies to science. Elites *that are isolated from politics* are a condition for the survival of democracy, acting as a check and balance on would-be dictators. Replacing these elites with more democratic institutions does not work: the bet on the bottom turtle of truth no longer makes sense. The revolt against MMR vaccine is a good example of how things go wrong with COVID vaccination as a more recent example.

Vaccination shows that science is integral even to the more extreme concepts of freedom because my freedom ends at the end of your nose and, in the case of vaccination, it takes science to explain where the end of your nose is. Climate change shows why even your distant neighbours, distant in space and time (and that includes future generations), have noses that belong in the calculation. At the same time, some parties, such as conspiracy theorists, have an interest in hiding the end of others' noses, whether they belong to the young or sick exposed to disease because of the absence of herd immunity, or, in the case of industrialists, citizens who die of lung cancer or distant strangers who are exposed to rising sea levels.

The new course would explain how the integrity of science is a role model for democracy and show how science's values and aspirations can be even more important than science's findings because of the way they infect society with truth. The course would explain what veritocracy means and how polysemic sloganizing works and how it was used in the Brexit debate and in the growth of neoliberalism and subsequent erosion of truth. Included would be an explanation of why the citizenry need such a course: it is because of the law of conservation of democracy which requires that citizens understand the role of truth in society and where truth is to be located.[30]

Finally, if the discussion were taken up by the media, then it might also reach the family home and become a part of early socialization. For this to happen citizens need to be encouraged to engage with the political system as part of their everyday lives, which is a challenge at the moment given current trends. For example, a recent IPPR report shows that voter turnout has been falling across Europe since the 1950s. Somewhat ironically the exception is the UK, where voter turnout has increased over the past decade, albeit from a very low base that means it remains below the average for advanced economies. The same report also documents high levels of dissatisfaction with politics: when people were asked whether 'democracy serves them well', the young and the most deprived were the least likely to agree.

If, however, a sense might grow that the outcome of elections might make a meaningful difference, perhaps to the extent that the government would rule *as a veritocracy*, then maybe things would change. Something like this seems to have happened in Poland where, in the recent elections, turnout was over 70 per cent the proportion of 18- to 29-year-olds who voted exceeded that for the over 60s.

If discussion did reach into the home, it would be the best way for veritocratic thinking to become integrated into citizens' ubiquitous expertise; in the home socialization starts early. As a fall back it should be taught near the beginning of secondary socialization. Promulgating veritocracy – a respect for truth – is at the heart of the preservation of a democracy that does not too easily descend into populism and fascism.

Column 5: A new political culture

Column 5 represents the top level of the fractal model and is another way of restating our core argument: that politics is downstream of culture. The bottom of Column 5 is the state we want to reach – a pluralist democracy robust enough in virtue of its culture to survive in the face of the law of conservation of democracy and robust enough to avoid the descent into authoritarian populism (or worse). A crucial mechanism would be primary socialization and civic education. The middle of Column 5 is a summary of what this cultural change would look like: truth-telling and integrity would be found at the level of ubiquitous expertise and, therefore, would be politically attractive enough to figure on political manifestos.

For this to happen, cultural changes are needed but they can be built on what is already there. As the public response to Liz Truss's minibudget, and to what are seen as Boris Johnson's lies, and the relative failure of Trump-backed candidates in the US mid-term elections show, there are still those amongst the voting public who want pluralist democracy to survive.[31] What veritocracy would need to add to this is that, as part of doing democracy successfully, the same public is also open to persuasion that science is the best way to begin the approach to questions of policy in respect of advice about the observable world. Our hope is that, not only would they accept the importance of science as a check and balance on the claims politicians can make about the social and natural worlds, but that citizens would also see that its moral commitment to the truth provides an even more important service to democracy.

The role of science summarized

The above aspirations all turn on a view of science that has been put together throughout the book. Figure 7.1 summarizes its various features.

The large stick figure is there to remind us that humans are not the free choice-makers that neoliberal economics describes but are made up of social groups into which they have been socialized. But it is important that the top-level ubiquitous group is universal across human societies: there cannot be languages without truth and consistency. As we go down the fractal the kinds of socialization that are available vary from society to society and institution to institution within any one society. But a society is its sub groups – that is where its population is located – so its sub groups can influence its ubiquitous expertises. Truth once established in early socialization can be bolstered or eroded in later socialization.

We've argued that it is erosion that is the major force in current Western politics. Science, we say, has the potential to turn it around. It can underpin the basic ubiquitous moral code because of the way it works as an institution.

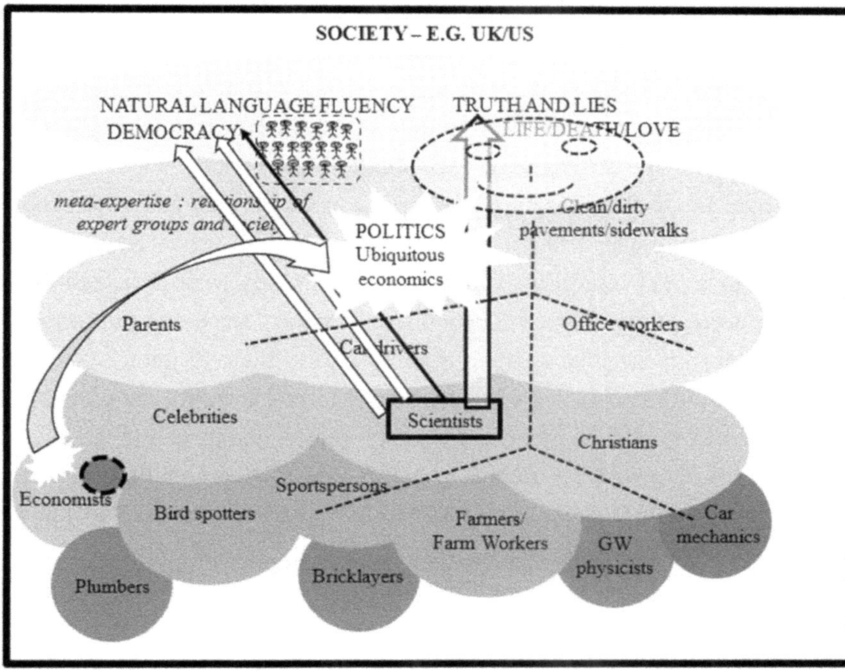

Figure 7.1 The elements of the argument (created by Harry Collins).

It seeks correspondence truth and depends on moral truth. That is represented by the prominent vertical arrow in Figure 7.1. Science also acts as an object lesson for decision-making in democracy because it has, continually, to make decisions under conditions of uncertainty while cleaving to moral truth: that is represented by one of the two, narrower, white, left-sloping arrows. Also, science, like the law, and other features of Western democracies, acts as a check and balance, countering the whim of any would be dictator; that is the other of the two narrower, white, left-sloping arrows.

We are talking here of the formative elements of science as an institution. Much of economics (Chapter 6) doesn't fit this positive model very well. There is disagreement in academic economics of the kind to be expected of a truth-seeking social science, but in the formative years following the Second World War and the environment of the Cold War, 'heterodox economics' struggled to make an impact even in the universities. The influence of big business via its support of academic appointments and politically inspired think tanks is described in Chapter 6; this impact transferred itself via an obsession with markets (the white ragged star) into academic economics and into public and political debate. The result has been a 'ubiquitous science' making economics the plaything of the ordinary public, which, as a science, it should never be.

Also, academic economics and ubiquitous economics still show the scars of a mistaken image of what a science should be – a mathematically based, physics-like, set of theorems and calculations based on a simplified model of humanity (*homo economicus* consisting of an aggregate of free choices), and this leads to model reification which pulls economics away from assiduous search for correspondence truth. Even if contemporary teaching in universities is more open to alternative views, it is still pulled to the reification of models. Nevertheless, in the public sphere (the white ragged star within the ubiquitous expertise domain labelled 'politics and ubiquitous economics' and fed by the long white curved arrow) economics continues to be misrepresented as a science which supports economic policies (neoliberalism. market fundamentalism), for which – 'TINA' – there is no alternative.

The narrow black, left-sloping arrow is far more part of regular political philosophy: without science we would not know how to find 'the end of our noses' so we would not know how to balance our personal freedom to act on the one hand, with, on the other, the damage that such freedom causes to our fellows – who are represented by the stick figures in the small rectangle. Science does not provide an answer to the proper balance, but it does provide the data on which the decision should be based.

There is no tension between technical experts and democracy – on the contrary, trusted technical experts are needed to check and balance political power. In a veritocracy, these things would be part of the taken-for-granted understanding of citizens.

What is veritocracy once more

'Veritocracy' is a neologism because there was no need for such a term in decades past. In the childhoods of the oldest among us, it was expected that palpably dishonest politicians would not attract votes, and it was expected that politicians caught lying in the course of parliamentary business would resign. But the antics of Trump and Johnson and their cronies have made it that this is no longer taken for granted in the West. It is true that Johnson was forced out by his parliamentary opponents for appearing to lie, but that was only after he had been perceived as telling far more lies than most British people believed a prime minister could survive. And even after he left, the strongest candidate to be his successor was one of his closest allies who had refused to take part in the revolt against him or criticize his attacks on truth – someone who one might have thought would be automatically disqualified from the leadership by those facts.[32] In the United States, Trump continues to deploy the claim about the election result to win nomination for himself and his favoured candidates and, sadly, he now exhibits the working out of the law of conservation of democracy.

A veritocracy is a democracy in which the idea of the value of truth has become so entrenched in the general culture that those mounting political campaigns would find it attractive to voters to include a promise to tell the truth; the value of truth would become still more attractive as it becomes reinforced by the positive feedback of campaign affirmations. Polysemic slogans would no longer be attractive to voters because they would want to know the truths that underlie political slogans. Under these circumstances, honesty among politicians would grow to become a component in what economists call the 'utility functions' of ordinary citizens because honesty will, once more, be a formative component of the language. When we learn a natural language, along with it we learn a moral code and people learning the term 'veritocracy', with its associated network of related terms, will be learning that the citizens of veritocratic countries value honesty above the many other things they might desire; they will come to feel the warmth of honesty and the chill of lies – these contrasting sensations will feed into citizens' utility functions. That is how we hope the culture will change, and remember, politics is downstream of culture. The law of conservation of democracy is just one expression of it.

In a veritocracy, all of this could happen. It is not an impossible dream: we have seen veritocrats in the past. Jimmy Carter's campaign theme promised:

> 'a government as good and as honest and as decent and as competent and as compassionate and as filled with love as are the American people,' … he [would] restore the trust of the people in their government.
>
> (Krukones 1985: 137)

And the current leader of the opposition in the UK, the Labour Party's Keir Starmer, reacting to Johnson, has recently promised: 'What you will always get from me is someone who believes honesty and integrity matter.'[33] Furthermore, part way through writing this book, Rishi Sunak, the UK successor to the chaotic regimes of Prime Ministers Johnson and Truss, though a member of the same Conservative Party, campaigned successfully in October 2022, to govern with 'integrity, professionalism and accountability'. Is this the beginning of veritocracy in Britain? Words that connote veritocracy have already seen as attractive in political campaigns in US and UK politics. Admittedly, the Carter example goes back a long way.

Our hope is that a combination of these arguments and these interpretations of what has been observed will bring science back into public esteem, make it attractive for politicians to endorse honesty and integrity, and create a series of positive feedback loops that will change culture in a direction that prevents the erosion of truth in the West.

But isn't this simply naïve utopianism

Perhaps the state we are in is the outcome of material forces that no amount of academic argument can change. Trying to change the way a population thinks may be futile in the absence of changes for better in the way they live. There is a debilitating cynicism that comes with the social scientist's tendency to see social changes only in terms of large-scale trends; it is disempowering. But politicians of the right, exemplified by Fisher's conversation with Hayek, do not seem to think that interventions at the level of culture are futile: on the contrary, they pour resources and energy into changing our ways of thinking from academic economics and think tanks to Fox News while resolutely keeping the poor poor, or even poorer. Why does the right, with Trump as the icon, vigorously exploit the new technologies of culture change: social media and, just around the corner, artificially 'intelligent' conversationalists who lack only the vote when it comes to political influence on our culture (and an individual vote is the least potent of politically efficacious actions)?[34] Why should those who are more left-inclined prefer a cynical shrug of the shoulders, particularly at a time like this when our future hangs in the balance. Perhaps the forces we confront are too large and too material for us to affect. But those with access to the means of cultural change do not act as though that were the case. They act, as we should act – as though we might be triggering the crucial flap of the butterfly's wing. The 2024 election in the United States might be that wing: a flap and the West has collapsed into fascism for ever. If we are already too late, we had better be ready to flap again and hope for the best.

What we have tried to do here is use what we know about science and about culture and the way they do or should interact to counter the current slide into fascism in the West. Like dutiful academics in the realms of science or philosophy, we talk in the register of eternal verities – literally in this case. But we must consider that this talk might also be the trigger in a chaotic system. We believe that any academic whose work bears on the issues should be considering today's moment and acting accordingly. Maybe this is futile and smiling cynicism is the more sophisticated attitude: time will tell and, if it all goes wrong, we can all tell our grandchildren, if they survive, what we did at the time of the great watershed. Let's hope the cynics are justified and doing nothing was OK all along.

Notes

Introduction

1 Levitsky and Ziblatt's (2023), *The Tyranny of the Minority*, takes us through the way the existence of slavery distorted the American Constitution, the way, even after the Civil War, democracy in the Southern states was vitiated by violence at the polling stations, and how fear of a multi-racial society is stultifying change to this day, but they also catalogue the repeated failure of multiple attempts to change the Constitution and change the rules of procedure in a way that would allow democracy to be reinforced. They also show how change via the polling stations is made much more difficult because the rules allow for minority government at all levels so the reaction to undemocratic behaviour has to be more than marginal if it is to be stopped at the polls. See also their 2018 and see Turchin (2023). Turchin is an analyst of material factors concentrating on wealth inequality. He claims that to avoid disaster we need political reforms that will moderate the 'wealth pump' in Western societies that has been in place since the 1970s development of the neoliberal consensus. The wealth pump is responsible for increasing immiseration and polarization notably via a great increase in the very wealthy and an overproduction of elites many of whom are inevitably starved of power. His thesis fits well with our analysis of the development of economics in the West.

2 It seems that 'veritocracy' is not quite the neologism that we thought it was throughout the writing of all but the final drafts of this book. But we are sticking with our definition (see sub-heading 'Introducing veritocracy', in Chapter 1). The previous uses, see https://internationaltimes.it/veritocracy-casting-a-mould-for-enlightened-governance/ and https://variety.com/2009/digital/columns/veritocracy-limits-online-exposure-1118001483/, seem rare enough for the term not to appear in dictionaries.

3 Most of this book and these passages in particular were written before the second Trump era; it is terrifying to see how soon so much of what we wrote has come to pass.

4 Wendy Brown (2019), for example, describes the material and economic conditions that are favouring the decline of democracy in countries like the United States.

5 Levitsky and Ziblatt (2023), describe how Victor Orban has effectively destroyed the rule of law in Hungary by filling all legal vacancies with his followers. Kakutani

(2018), is a history of the decline of truth in the United States culminating in the rise of Trump and also the exploitation of postmodernism in Russian domestic and international politics.

Chapter 1

1 See Collins and Evans (2017b). We are all going to have to accept that science is not foolproof, and we don't know how to make it foolproof; that kind of security for scientific knowledge cannot be found in today's philosophical world. It remains, however, that we spend most of our time relying on scientific knowledge and mostly this is a sound judgement. Here we are trying to explain why: it is not that science is perfect but that the people who make science are unusually obsessed with truth, and if you want something material to rely on it's better to rely on people who are obsessed with correspondence truth than those who are obsessed with money, or markets, or political success, or this religion or that. To change the minds of populists in this regard means changing the culture and this is going to be difficult; in the final chapter we will suggest that the schoolroom is going to be vital.

2 For a recent re-print of the work, first published in English in 1958, see Durkheim (2019).

3 For the concept of 'formative intention' or 'formative aspiration', see Collins and Kusch (1998). A formative intention is not a norm but an aiming point. A norm is a uniform way to act, whereas an aiming point is something that leads on to the choice of norms. When we get to Table 1.2, we'll find the norms of science in the third column. We turn Merton's analysis of science on its head: Merton believed scientific truth arose out of democracy because of the alignment of democracy's norms with the norms of science; we start with truth and that leads us on to something resembling Merton's norms and the use of science to support democracy. For more on the socio-political context in which Merton's early work on the ethos of science emerged, see David Hollinger's (1996) essay on 'The Defence of Democracy and Robert K Merton's Formulation of the Scientific Ethos'.

4 One source for the deeper philosophical point is Nelson Goodman's (1955/1973) solution to his 'new riddle of induction', which links consistency of observation and language: he said our inductive choices, such as whether to see grass as green and the sky as blue, are formed by the degree of 'entrenchment' of such colour terms rather than other colour terms in the language. Goodman's argument was taken up in Collins's (1985/1992) and applied to science; Collins points out that without

constancy of usage – life would be impossible. For instance, without a language that embodies consistency we would never dare get on a bus because we would be expecting it to turn into something else (Ch 1 p9). Truth can also be thought of as a foundation to the idea of interactional expertise (Collins 2004a; Collins and Evans 2007). Interactional expertise is fluency in a technical language such as to allow its possessor to make sound technical judgements even though they may not be able to execute them in practice (which requires contributory expertise). The potency of language has another empirical counterpart in the study of word-embedding in language corpuses as practised in artificial intelligence (described in, for example, Collins 2018). Since this book is about national cultures, we'll occasionally draw on a variety of sources whose job is to explore and report on culture, notably contemporary culture: journalists and the broadcast media, documentary makers, notably Adam Curtis, and, in one instance, a historically inspired novelist, Amitav Ghosh.

5 Sometimes this is discussed in terms of the 'Overton Window' of political possibilities, named after American policy analyst, Joseph Overton.

6 Durkheim, in his book, is always anxious to back up his claims about anomie with statistics: the statistics of murders, thefts and the like in various countries. There are no statistics about truth and lies, however, only the sense gained through immersion in the society, which is why our sources of evidence are what they are. But this does not make these sources as fragile as they are said to be when compared with the numerical sciences (for a discussion see Collins 2019; Collins and Evans 2017a).

7 Formative aspirations differ from Mertonian norms in that they are both more and less deterministic. They are more deterministic in the sense that, if there is no desire to find correspondence truth, then science is not being done. They are more flexible in the sense that they say very little about the methods by which this goal is to be achieved. In contrast, Merton's norms, at least for his critics, seem less flexible, with the existence of counter-norms taken to invalidate their use as demarcation criteria rather than demonstrating that the activity was no longer a science.

8 For some ideas along these lines, see Collins, Bartlett and Reyes-Galindo (2017).

9 For discussion pertaining to each of the numbered definitions see:
1: The popular science writings of Richard Dawkins (e.g. Dawkins 1976, 1986) are amongst the most extreme version of this view. Others, such as Carlo Ravelli (e.g. Rovelli 2021), provide a less imperialistic version of the same basic idea.
2: Steven Shapin's (2016) 'Invisible Science' is a good example of this way of thinking about science. See also Brian Balmer's (2012) *Secrecy and Science*, which documents the UK's biological and chemical weapons research programmes during the post-war period.

3: A good survey of the dangers that are faced by the institution of science as it grows is Turner and Chubin (2020). The cultural core of science has to be preserved from these predatory changes.
4: This is the view that animated early STS work such as the strong programme (Barnes, Bloor and Henry 1996; Bloor 1991; MacKenzie 1981) or the 'Bath School' (Collins 1974, 1981, 1985/1992).
5: This is the semiotic tradition that developed within STS. Actor-Network Theory is the most famous expression of this approach (Latour 1993, 2005; Latour and Woolgar 1979).
6: This is the co-productionist view that sees science is inevitably intwined in the work of the state and often co-opted by powerful interests within those institutions (Jasanoff 2004).
7: This is the version of science described by SEE and which is the foundation of the argument in this book. For the origin, see Collins and Evans (2002); for other statements of this position, see Collins and Evans (2017b) and Collins et al. (2020).

10 For the Russian case, and especially the role of Putin and advisors such as Vladislav Surkov, see Pomerantsev (2014, 2019). And for the obviousness of what seems to be taking people by surprise, see Turchin (2023).

11 Collins's 45-year-long immersion in the field of gravitational wave detection was part of the inspiration for this description of science (see, e.g. Collins 2004b, 2013, 2017). The seekers after gravitational waves fitted the aspirations of a true science remarkably closely with only a handful of members of that thousand-plus strong community being ready to sacrifice truth for personal interest (though Collins complains about the desire for secrecy leading to lying in his 2017 book describing the first discovery). We can also see the difference between institutions like physics and commercially orientated institutions like artificial intelligence (AI) research. A science like physics is paranoid about claiming a false discovery: it happens from time to time but the promulgators are vilified and physics often assembles 'red teams' to criticize fiercely before anything is released to the public domain. AI's motto, on the other hand, is 'move fast and break things'. And, indeed, it continually breaks it promises – for example, it continually claims to have developed machines that have passed the Turing Test – ridiculous claims when the Test is treated as something other than a game.

12 It gets complicated with deceits like the 'palter'. For example, 'I did not break the lamp last Friday', might be a true claim but deceptive as the questioner just wants to know whether you broke the lamp. Likewise, with deceits of omission where an incorrect understanding is knowingly left unrectified. But promulgators of palters, and deceits of omission, aren't mistaking themselves for truth-tellers, they know they are lying but are playing games with words. Moral truth is interesting in that it is hard to imagine someone who intends to tell the truth but accidentally

lies because they can't tell the difference between truth and lies. This can easily be the case where correspondence truth is concerned but should someone really not be able to tell the difference between intending to lie and intending to tell the truth, we enter the mysterious realm alluded to by Hannah Arendt in the alluded to by Hannah Arendt in the text quoted at the start of the Introduction, the world of Kellyanne Conway, Trump and Putin, or perhaps the realm of pathological psychology. As we will go on to explain, one can be mistaken about whether one has satisfied one's intention to execute moral truth, not only because one might be sincerely wrong about the world but also because one's attempt to tell the truth is vitiated by unknown biases; that is why we have double-blind experiments and the like in science, but this is not the same as being mistaken about whether one was *trying* to tell the truth or not.

13 Here is an almost morally legitimate exception to this generality. What is now agreed to be the first detected gravitational wave impacted on the LIGO detectors on 14 September 2015, but before the scientists were ready to announce it publicly (which happened in February 2016). The scientific community (in which Collins was embedded) needed to do several months work to confirm their initial findings and eliminate possible mistakes. They decided to keep their proto-discovery secret during this time. As a duty to his scientist-hosts Collins had to conceal his knowledge too. For example, he once said to a reporter who was questioning him that he had not heard the discovery rumours the reporter was relating – which, strictly, he hadn't – whereas Collins knew the reporter really wanted to know whether the rumours were true – which he knew they were (a palter). These events are written up in Collins's (2017), *Gravity's Kiss*.

14 The importance of defending science on the basis of its values rather than the expertise of its practitioners can be illustrated by the following episode in the recent history of science. In the 1970s the Israeli showman Uri Geller burst onto the scene. Geller claimed to be able to do various paranormal things that certain scientists who saw them declared to be beyond the understanding of science. Over a period, certain stage magicians claimed they could debunk Geller and could repeat his tricks using subterfuge 'under laboratory conditions', but that only they understood enough of the techniques to be able to do it. A number of scientists, and other prominent representatives of science, such as the editor of *Nature*, endorsed the stage magicians' new role in science. Here were scientists bringing in an outside body of professionals to help them establish correspondence truth. All of these were experts in some sense and in some domain but none, we argue, could be seen as doing science as they worked with an entirely different set of values: for example, they refused to explain their methods and techniques, referring only to their

authority as skilled stage magicians, they refused to accept certain invitations to repeat Geller-type performances under laboratory conditions specifically designed to reveal the trickery, and their profession of stage magic was based on deceit. Therefore, according to our analysis one would have been ill-advised to bet on their advice irrespective of the actual truth of the matter.

Another way of framing the same analysis is to see it as something like the problem of fringe science. The existence of interpretative flexibility means that it is impossible to apply scientific methods to defeat every fringe proposal or science would come to a halt, so scientists use 'social criteria' to reject these outsiders. Here, faced with a fringe or pseudo-science, the mainstream scientists allowed outsiders without scientific skills or norms to do the 'dirty work' of rejection for them. The bad thing was that the spokespersons for science presented what was going on as if it were real science, whereas it was theatre and showmanship. The presentation of what was going on encouraged the general public to misunderstand the nature of science even though the net outcome probably corresponded to the way the world is.

15 Collins et al. (2022), *The Face-to-Face Principle* analyses the importance of face-to-face communication in science and the negative impact of social media upon science's role in contemporary politics. (The book is available on open access.)

16 Collins (2001) describes the events.

17 The incident is described in Collins (2017) *Gravity's Kiss*.

18 Which, incidentally, means that the current enthusiasm for 'open science' is worryingly misplaced. See also Collins (2014).

Chapter 2

1 The Three Wave schema was first put forward in Collins and Evans (2002). The paper also introduced a new analysis of expertise which is developed in Collins and Evans (2007). (For an alternative approach to the analysis of expertise in society, see Eyal 2019.) It requires a charitable approach to recognize a schema of this sort because there are always precursors and thinkers who do not fit neatly into the temporal epoch assigned to them. If the epochs are relatively contemporary it is also galling to find one's agonisingly established ideas re-assessed as merely part of a trend (though sociologists and historians ought to be ready to for it). Whatever, when first published, the Three-Wave schema experienced much opposition, some intemperate. In part that opposition was invited because it was posing an alternative to the popular trend towards democratization of science. Subsequently, the original paper and its follow-ups have been very highly cited

and its proposed analysis of expertise much used and widely promulgated, in handbooks and the like, in psychology and philosophy, though less frequently in science and technology studies.

2 It would be tedious to try to list a bibliography of the First Wave and the literature is very well known. Popper brought up the rear, as we might say, and symbolized the intention with his (1959) *Logic of Scientific Discovery*. Kuhn's (1962) book perhaps marked the start of the shift to Wave Two and some people spoke of a Popper-Kuhn debate but it was later work that gave rise to the change in academic culture, with Bloor's (1973) paper probably being the first publication marking the potential 'paradigm shift'.

3 Winch (1958) makes the connection from the philosophical side of the argument whilst Bloor's *Knowledge and Social Imagery*, first published in 1976, makes the opposite journey.

4 See Scheffler (1967: 81). Sadly, the reception by scientists of the amusing but philosophically shallow 'Sokal hoax' attack on literary analysts of science was, perhaps, a better example of mob psychology. For a discussion of the Sokal hoax in the context of hoaxes in general, see Collins (2008).

5 Kuhn (1962), Wittgenstein (1953). The exposition of Wittgenstein provided by Winch (1958) was important to some analysts. A Wave Two analyst *avant la lettre* and an anticipator of much of what was said by Kuhn is Fleck (1935/1979). Pierre Duhem's (1908/1981) understanding of the way scientific facts are embedded in networks of assumptions is still important to this day but there are a number of other precursors.

6 Bruno Latour, by bringing the literary analysis into the equation, was responsible for a huge expansion of science studies into the humanities. He argued that science could be analysed in ways that did not depend on any understanding of technicalities of the science, thus giving humanities scholars new weapons in the war between what CP Snow called *The Two Cultures*. It can be argued that the semiotic approach was essentially conservative, however, since it took creative agency away from humans, where it had come to be located in the sociology of scientific knowledge, and gave it back to the material world where it had always been located in scientific texts.

7 It was Kuhn who remarked on how the textbooks mythologized scientific history and this was confirmed in *The Golem* series of studies, with the second edition of the first volume (Collins and Pinch 1993/1998), having considerable impact. For instance, the second edition showed how inaccurately a haphazard selection of physics texts, not to mention Stephen Hawking's (1988) *Brief History of Time* reported on the Michelson-Morley experiment.

8 This point should not be confused with the 'reproducibility crisis' which grows out of too optimistic an understanding of the power of statistics in science or

the misuse of statistics. The experimenter's regress applies to all pioneering experiments, including physics discoveries claiming a 5-sigma significance level (intended to swamp unknown systematic errors, the point being that they remain unknown until the new science is fully sedimented). In the light of physicists' experience, sciences using a 2- or 3-sigma criterion were always going to be in trouble however well-intentioned the experimenters and they are hopeless when the experimenters are careless or corrupt. The point is not unrelated to but is not identical to Duhem's idea that all theoretical claims in physics rest on bodies of assumptions – Duhem had no concept of the nature of experimental skill. We will use Duhem's basic idea in various places, however, and the two ideas merge if we allow that trust in an experimenter's ability is an assumption that experimenters must make when they utilize another's results.

9 For example, McCright (2016) shows that climate change scepticism, even amongst 'regular citizens' is correlated with trust in, or identification with, the industrial capitalist system. These citizens also tend to support policies and parties that have a similar stance. For those who accept the reality of climate change, the opposite is the case: trust in, and identification with, groups that challenge this system is a key predictor of beliefs, policy preferences and voting intentions.

10 As remarked in an earlier footnote, all this is explained at length in the second edition of Collins and Pinch's (1983/8) *The Golem* second edition of Collins and Pinch's (1983/8) The Golem: see Chapter 2 and the new 'Afterword'. For an example of how the faux exactness of science can be used by its opponents, see Jordan Peterson's rant against the science of climate change in *The Telegraph*, which can be found at Jordan Peterson: Peddlers of Environmental Doom Have Shown Their True Totalitarian Colors – Watts up With That? repeated in his blog (Article: Back Off, Oh Masters of the Universe): Peterson seems to think that failure to make exact predictions are failures of climate science and this gives him licence to insist that the future of the climate must be left to market forces alone.

11 Source: Jasanoff (2013: 101).

12 An early intimation of the debate to come is Scott, Richards and Martin (1990) 'Captives of Controversy: The Myth of the Neutral Social Researcher in Contemporary Scientific Controversies', and Collins's (1996) response entitled 'In Praise of Futile Gestures'.

13 See Mirowski (2020) for a detailed historical discussion of this movement and its contribution to contemporary politics. Mirowski is, if anything, less forgiving of the democratizers. He writes:

> Harry Collins has recently suggested that 'Contemporary STS erodes the cultural importance of scientific expertise and unwittingly supports the rise of populism' (Collins et al. 2020: 1). By contrast, I believe the Leveller faction [Mirowski's term for what we call the democratisers] has developed its political stance in a slightly

> less roundabout and absent-minded fashion than he lets on. One needs to ask: What exactly has STS done to oppose or otherwise counter the rise of post-truth practices? … the fact of the matter is that Levellers have by and large supported the kinds of knowledge structures that enable the advent of the post-truth era and populist movements. They have [denied] … any substantive distinctions between science and politics.
>
> (p25)

14 See e.g. Dewey (1927/1954); Lippmann (1925); Schudson (2008). See Collins and Evans (2017b: 112ff) for a summary of the debate.

15 From a 2022 presentation at Cambridge University entitled 'Democracy and distrust after the pandemic' From a 2022 presentation at Cambridge University entitled 'Democracy and distrust after the pandemic'. Darrin Durant Darrin Durant (in preparation 'What did Kant say about the individual and trust in expertise?') suggests Jasanoff has misunderstood Kant, attributing to Kant an individualistic valorization of solitary reasoning and thereby overlooking Kant's focus on our epistemic interdependencies as the path to enlightenment. Durant claims Jasanoff makes four errors. First, Jasanoff tears Kant's maxim 'think for yourself' out of context, making it about autonomy through self-determined communication. Yet for Kant 'thinking for yourself' is 'a maxim of unprejudiced thought' (WOT, VIII, 145; 303), demanding communication be more than an echo and instead a plurality of viewpoints aired in conversation. Second, Jasanoff further misrepresents Kant's 'think for yourself', in her examples of COVID-19 public health restrictions and guidelines, as a presumptive act of distrust. Yet Durant shows for Kant it is the opposite. Kant outlines a presumptive acceptance of the testimony of others: '[W]e can just as well accept something on the testimony of others as on our own experience' (AA, Vol. XXIV.2 (*Vienna Logic*), 896). Third, Jasanoff misrepresents Kant as suggesting a social compact in which citizens distrust experts and thereby hold them to account, but Durant suggests Kant was suspicious of suspicion. Kant criticized two forms of 'incredulity', either '[accepting] nothing on testimony [unless] it is sufficiently confirmed to be knowledge' or 'not want[ing] to accept anything as true except on theoretically conclusive grounds' (AA, Vol. XVI (*Logic Nachlaß*), 508). Durant discusses the modern version of these two problems: doubt mongering about expert observations and questing for absolute certainty to delay regulatory action. Fourth, contra Jasanoff, for Kant the distrusting of expert testimony is itself a moral problem, because incredulity is a 'lack of moral interest' (AA, Vol. XVI (*Logic Nachlaß*), 509) in the epistemic interdependencies at the heart of reasoning in public. Incredulity, said Kant, is 'logical egoism': '(Logical egoism is a selfish prejudice.) This is not merely self-conceit, but a kind of logical principle, which deems superfluous the criterion of truth: to compare

one's opinions with those of other people ... The principle of indifference etc. to the judgments of others in comparison with my own is the principle of logical egoism' (AA, Vol. XXIV.2 (*Dohna-Wundlacken Logic*), 740). Logical egoism breeds a moral problem because speakers collude with gullible or self-interested audiences 'to make imperceptible the blatantly obvious inequality between loquacious ignorance and thorough science' (AA, Vol. XI (*Correspondence II*), 141). Falsely assuming epistemic parity in cases of conflicting testimony is morally blameworthy because it involves demanding that 'others think little of themselves in comparison with us' (AA, Vol. VI (*Metaphysics of Morals*), 465). Fifth, Kant specified a scope to 'think for yourself', aiming it at moral truths and claims which are a prior (analytic). But testimony about contingent, fallible experiences – what we might call expert testimony (Kant spoke of scholars, history and science – experimental physics, geography) – presented a presumptive obligation to rely on other's experiences (those in a better position to know) (AA, Vol. XXIV.1 (*Blomberg Logic*)). Overall, Durant suggests scholarship on Kantian testimony does not support Jasanoff's mobilization of Kant to suggest modernity's delegation to experts is a fall from grace vis-à-vis 'thinking for oneself'.

16 See, e.g. the report by Philippa Roxby for BBC News (Roxby 2024).

17 E.g. see Wynne (1989, 1992) in which the claims were widely promulgated. The 1992 paper has been especially influential and has nearly 3,000 citations according to google scholar.

18 Collins and Evans (2002) paper discusses the Third Wave in terms of the tension between the problem of legitimacy – the acceptance of scientific expertise by the public – and the problem of extension. For the authors' point of view with references to other viewpoints, see Collins, Evans and Weinel (2017).

19 There is a purely professional interest that bears on the success of the democratizers within the academic sphere. On the whole, the potential influence of the public is easily demonstrated in fieldwork involving small local issues. Such studies are ideal subjects for PhD work. Therefore, in the social studies of science, small studies are very much favoured. Studies of big questions to do with the nature of science, and which would not point to the existence of any expertise among the public, are not so easy to carry out.

20 Collins and Evans (2007), includes a 'periodic table of expertises' that classifies the various kinds of expertise from ubiquitous expertise of the public down to specialist expertises.

21 Something that can be linked back to Goodman's new riddle of induction as discussed in note 4 in Chapter 1.

22 This finds a parallel in the questioning of the use of expert witnesses in the law, where technical expertise is involved and which is to be understood by lay juries. The argument made by Caudill, Collins and Evans (2024) is that the technical

decisions should be left to the technical experts, perhaps in consultation with the judge, though the overall decision is still the prerogative of the jury.

23 See Elliot (2022); Douglas (2021); Steel (2021).

24 See Collins and Evans (2017b, Table 2.2., p. 55), which includes a more elaborated set of values internal to science.

Chapter 3

1 A topic to which we will return in Chapter 6.

2 We capitalize 'Imitation Game' to refer to the version with a carefully worked out protocol used in various experiments designed by the Cardiff University group but the philosophical points apply in principle to *the concept* of an imitation game.

3 The distinctiveness of the expertise of office-workers was, perhaps, more distinctive a couple of generations back.

4 In more technical language, the solution desired by the democratizers is to change the 'locus of legitimate interpretation' (LLI) from that traditional in the sciences to that more appropriate for the adventurous and performing arts or for business enterprises. Jumping ahead somewhat, Mirowski (forthcoming) suggests that the current enthusiasm for 'open science' is associated with neoliberalism because scientific outcomes are best thought of as emerging from free markets. It could be that this is also seen as corresponding to the Mertonian (1942) norm of 'communism' which has to do with the free sharing of knowledge among scientists and might, in turn, be thought of as related to the norm of clarity which facilitates productive criticism. But these norms work well only within the restricted and narrow domains of specialists – they are hopelessly misplaced when taken to apply to science removed from narrow domains, as the LLI indicates.

Chapter 4

1 The concept of 'inductive risk' is one of the reasons why it not always crystal clear that the public cannot have an influence even at the research front. If choices made at the research front, such as whether a certain level of statistical significance justifies the announcement of a result, depends on the possible consequences of that result for policy action, then the public are indirectly affecting science at the research front, but not in the way that the democratizers advertise.

2 Continue to bear in mind also that much of what is found in the profession of science is not proper scientific expertise.

3 See Collins et al. (2022).
4 See Collins (1974, 1985/1992).
5 The over-reading of the value of the 2-sigma level is found in most sciences and cannot be resolved by meta-analysis because of the tacit-laden nature and systematic errors found in of difficult experiments.
6 To borrow the language from another classification, we might say that the lower half of the table corresponds to what Funtowicz and Ravetz (1993) call 'applied science', by which they mean settings where the uncertainties and decision-stakes are sufficiently low that standard scientific methods are all that is needed, whilst the top half represents the move towards post-normal science, in which uncertainties and stakes increase, requiring a different approach to be taken. We recognize the change in type of science that this transition brings but differ in how we see its resolution. In particular, post-normal science leans much more towards the democratization of science than the approach taken here.
7 An extended field study of fringe physicists is Collins, Bartlett and Reyes-Galindo (2017).
8 But see also the concept of inductive risk discussed on page 45 and in note 1 in Chapter 4.
9 A recent paper by Morelock and Narita (2022) describes how conspiracy theories, such as those promoted by QAnon during the COVID-19 pandemic, not only question both political and epistemic legitimacy but also harness the power of social media and populist leaders to increase their profile and influence.
10 See Collins, Bartlett and Reyes-Galindo (2017) for an elaboration of this point and for a detailed field study of fringe physics and the range of more extreme fringe sciences.
11 Though it was disturbing, but not surprising, to discover that part of the challenge to Einsteinian physics at fringe physics conferences continues to be expressed in terms of antisemitism.
12 All this may change rapidly with the nightmare scenario growing fast as a result of the influence of Robert F. Kennedy Jr in the wake of second Trump election.
13 The information required is more like voting intentions than a uniform feature of the culture: a matter of surveys, not probes (Collins and Evans 2017a).
14 See, for example, Evans (2007).
15 For hyper-normal science, see Collins et al. (2023).

Chapter 5

1 As explained in the Preface, our prime focus is on the Western world though in this chapter we will range more widely. Reyes-Galindo's description of the relationship between science and government in contemporary Mexico is an object lesson in how science can fail to put a check on populist authority whatever the ambitions expressed here. He explains that scientific populism in the Global South and particularly Latin American settings does not fit the analysis of the United States and the United Kingdom. He says:

> Mexican science exists in such a precarious state of institutional dependence that it can hardly be an effective check on political power. In previous work I have illustrated how Mexican scientific institutions, conscious of this lack of power and dependency, rarely choose to confront and antagonize the state directly, even when there is clear ethical and epistemic obligation to do so.
>
> (Reyes-Galindo 2023: 223)

Although Galindo does not distinguish between upstream, downstream and research front intervention by the government, the 'epistemic obligation' reference suggests it is research front relevant. Ludwig and Poliseli (2018) provide a more general treatment of the relationship between traditional and academic knowledges.

2 This is a quotation from an official of the Ministry of Maori Affairs reported in Rip (2003) at page 421.

3 Galison (1997) provides the term trading zones, with Collins, Evans and Gorman (2007, 2017) analysing the problem in greater detail and introducing the ambassadorial model. These ideas were developed at a workshop organized by Mike Gorman and held in 2006 and published as (Gorman 2010). Our contribution was also published as Collins et al. (2007) and further developed to include the ambassadorial model in Collins et al. (2017).

4 They may be able to provide guidance in terms of 'second order measures of skill' (Pinch, Collins and Carbone 1996).

5 Standpoint epistemology sometimes links demographic diversity to epistemic benefit not via argument from particular cases but from the need, in science, for criticism. The argument then goes that the more diverse the criticism, the stronger will be the objectivity. But this is again subject to the problem of extension. Good criticism is informed criticism, not just more diverse criticism, and informed means 'cognitively narrow'. We will return to this point below.

6 As in an advertisement a few years ago for *Stella Artois* beer featuring French hop farmers.

7 The once much respected writings of Du Bois (1903/1994), who claimed, for example, that Black people had to be able to understand their white masters in order to serve them whereas whites had no obligation to understand their servants, are nowadays treated by many as no longer adequate thanks to the increasing influence of identity or 'body politics'.

8 For instance, a wide-ranging discussion by Walker (2022), who introduces the term 'body politics', covers among other things the attack on artist Dana Schutz for her sympathetic painting of the open-casket funeral of victim of lynching, Emmett Till. He puts a case very much aligned with what we are arguing here. Walker, however, does not mention interactional expertise and bases his analysis more on literature where imaginative empathy has a greater role than it does in social science.

9 One of the proximate triggers for this approach was the article by Abbot et al. (2023), who appear to want to re-start the science wars by taking diversity to be opposed to scientific merit. We share their belief that science should be strive to be as independent as possible, and that having quotas for types of persons co-authoring or being cited in books and papers is (a) impractical because one has to begin with a catalogue of types of person, the whole problem being made to seem too simple by the fact that women represent 50 per cent of the population everywhere and are uniformly oppressed (Harari 2015 makes the striking point that the groups selected for prejudicial treatment vary from culture to culture but women are oppressed wherever they are found, and we do not have a convincing theory to explain why), whereas other disadvantaged groups are represented in societies in a far more heterodox way; and (b) that, as explained above, solving the social justice problem does not solve the scientific problem. This very book-writing project was originally meant to have many more authors – six – of whom two were women. Unfortunately, the two women were the first to drop out so we soon had no diversity on the gender front. Does this make the project unsound scientifically? If it does, every small project is in danger. We do not think the new science-warrior approach is helpful, however, because of the oversimplification of the nature of science. We do not want to leave the relationship between science and the public clear for a futile war between extreme and ill-informed views.

10 In the Third Wave paper, we suggested that this responsibility should fall on the scientists.

11 These arguments do not apply once we move beyond the realm of the human because the key to virtual diversity is human-like language; animals cannot possess interactional expertise. Those social scientists and philosophers who emphasize the continuity between humans and animals should recognize that the basis of human societies is the extraordinary depth and richness of human language and that animal means of communication do not provide the same possibilities for wide

coordination of disparate actions and understandings across a range of specialist activities. Animals and other non-human entities do not possess human-like language and do not possess human-like culture. An attempt is made to prove, or, at least, significantly increase the credibility of, this position in Collins's (2020) paper entitled, 'Interactional Imogen: Language Practice and the Body'.

12 For interactional expertise lite, see Collins and Evans (2015).

13 The other two seem to be more about how the efforts to extend are received by the scientific community rather than prescriptions translatable into recommendations for scientific action.

14 As described in the field study by Collins, Bartlett and Galindo (2017).

15 The term 'problem of extension' was introduced in Collins and Evans's (2002) paper in *Social Studies of Science*, partly in response to the way the MMR revolt was being handled by prominent social scientists. The term 'co-production' is associated with what we have referred to elsewhere as the 'democratization' movement in STS – the move to make science more responsive to the citizenry. For what it is worth, at the time of writing in early 2024, the term 'co-production' had, since 2002, been mentioned in around ninety papers in *Social Studies of Science*, whereas the term 'problem of extension' had been mentioned in seven, which reduces to three once we remove ourselves and those who initially responded critically to the 2002 article (data collected 22 January 2024).

16 Thanks to Jacob Stegenga for pointing out the more recent interpretation. A popular account of it can be found here: https://www.sciencenews.org/article/rosalind-franklin-dna-structure-watson-crick.

17 The clear role of the contribution of social prejudice in the Franklin case is muddied by existence of similar cases among men alone. See, for example, Krige (2001).

18 That a ubiquitous expertise is relevant to a science does not mean that this is a ubiquitous science! That would only be the case if the population of men and women as a whole suddenly decided that since their ubiquitous expertises were relevant they were scientists by that fact.

19 This description is taken from the summary provided in Rosser (1992).

20 This example is described in Renck et al. (2022).

21 Wynne does not report the results of any interviews with MAFF scientists. As far as we can see, he investigated only the levels of trust and respect for the scientists among the farmers. Unlike in, say, a courtroom, as far as we can see, only one side of the disagreement is represented. Like in a courtroom, however, Wynne appears to take on the role of representing the sheep farmers rather than exploring cultural interaction. This is, perhaps, a symptom of starting the analysis from the point of view of (real or perceived) injustice rather than as a problem of the merging of

cultures, which would always involve symmetrical effort in respect of both sides (and see the discussion around Figure 1.2). We have approached Wynne (2 August 2023) in hopes of reviewing the case and any efforts to discuss matters with the MAFF scientists but up to the time of writing (May 2025) he has not responded.

22 Since 2002, in *Social Studies of Science*, the term 'MMR' has been mentioned seven times which reduces to four when we remove our own papers; none of those who initially defended the revolt against MMR appear in the author list of these four papers.

23 This example is based on the work of Martin Weinel (2007, 2010).

Chapter 6

1 For the source in Wittgenstein, see paragraphs 66 and 67 of *Philosophical Investigations* (Wittgenstein 1953).

2 Fourcade (2006) provides a fascinating analysis of the development of economics as a global profession. The paper describes how the emergence of mathematical modelling as the way to 'do' economics enabled its diffusion by uncoupling economic analysis from particular times and places but also how the United States and to a lesser extent the United Kingdom become key nodes in the network of accreditation and training that underpins the profession. Drawing on a range of reports, Fourcade includes statistics such as:

- In 1993, finance ministers in 9 out 12 'third world or former communist countries' had an economic degree, as did all but one of the central bank governors (166)
- In 1998, 79 out of 160 countries had at least one government leader with a degree in economics, including 16 with at least one leader with a Western PhD in economics (167)
- In 1999, economists recruited through the IMF's prestigious 'Economist' programme all graduated from North European or North American universities (179)
- In 2000, the majority of students on PhD programmes in both the United States (54 per cent) and the United Kingdom (approx. 90 per cent) were international students (172):

- The effect of this, and of the hierarchies within the economics profession, is perfectly captured in a 1999 interview with Arnold Harberger, a professor at Chicago University, which Fourcade quotes as follows:

I think my number of ministers is now crossing 25, and I know my number of central bank presidents has already crossed a dozen. Right now the central bank presidents of Chile, Argentina and Israel were my students, and the immediate former central bank presidents in Argentina, Chile and Costa Rica were also my students.

(180–1)

3 Though a look at Chapter 2 of Earle, Moran and Ward-Perkins's 2017, *Econocracy*, indicates, in a detailed way, that this, if it happens, is the exception not the rule. The general point about an education in economics can also be found in Michael Lewis's (1989), *Liar's Poker*.

4 Both authors know something about economics as well as other sciences and are able to compare them. Collins, as well as taking minor undergraduate courses in economics, spent more than twenty years in a School of Social Sciences that included the economics faculty, eventually becoming that School's head; in this period, he had endless discussion with economists about economics – effectively a socialization into its concepts and economists' ways of thinking. Evans has published a book about econometric modelling based on his PhD which was a sociology of economics. Both authors have reinforced their understanding to some degree via discussion with offspring who have undergone socialization into academic economics, one to degree level, and one to full professor level. We have also been able to discuss these matters during the course of writing this book with the various knowledgeable people mentioned in the acknowledgements, who might have been co-authors were it not for their commitments – some logistical and some political. Vitally important has been the published work of Oreskes and Conway on which we draw heavily for its documentation of the way academic economics has been influenced by society and the way ubiquitous economics (our word) has developed and influenced society: we were also fortunate in being able to discuss this work with Naomi Oreskes herself and with Deirdre McCloskey, one-time member of the Chicago School of economics who is now a Fellow of the Cato Institute think tank in Washington DC. For a more popular left-flavoured account of the rise of neoliberalism, see Monbiot and Hutchison (2024).

5 Unfortunately, Earle, Moran and Ward-Perkins's otherwise excellent (2017) book confuses the issue of public understanding of economics. The description on the back of the book claims 'Now economics is the supreme ideology of our time, with its own rules and language. The trouble is, most of us can't speak it'. But that's not the trouble; most of us can't speak the language of any of the specialist sciences whose advice we must draw upon. *The trouble* is that in the case of economics it is not really scientific advice that is being proffered and yet an awful

lot of ordinary people, such as the *Telegraph* readers addressed in the editorial of 23 September 2022 (see p150), think it is and think that they *can* speak the language of economics. Earle, Moran and Ward-Perkins's final chapter is a plea for a democratized economics with training classes for ordinary citizens taught to make economic/political judgements. Their 'thick democracy' seems to be the same as what Mirowski (forthcoming) criticizes, calling it 'Democracy2', which takes the democracy to mean something like what it meant in the Greek City State, with all members of the population continually engaged in the arguments whereas 'Democracy1' assumes the existence of expert politicians between whom the citizens choose. It just is not that easy, however, given the complexities of economics and given that the economic debate among professionals is a heated and largely political debate. The way forward has to be to get economists to change their practices, in the way the rest of this book recommends, thus making their discourse trustworthy *because* the technical uncertainties and political disagreements are revealed.

6 The 'mainstream' of the subject is sufficiently well-established to allow it to continue to refer to the critical wing as 'alternative economics' or 'heterodox economics' and to make it difficult for the critics to break away from that oppositional image and become part of a modified mainstream. The mainstream dominates in terms of relative numbers of practitioners, a more solid institutional foundation in university teaching, in outlets for publication in the top professional journals, which tend to exclude alternative approaches, and in opportunities for employment. Yet alternative or heterodox economics is not a fringe science because it is engaged in live academic debate with mainstream economists, and one could imagine it displacing much of the mainstream, especially if the quasi-market setting of university economics research, with its rigid hierarchy of journal outlets, were to change. (The contribution of Davis in Cartwright and Davis (2016) sets out the way the hierarchy of journals has made the subject resistant to change.) Heterodox economics is not driven by the pathological desire for novelty describable as an imbalance in respect of the essential tension, but a continual, though not very successful, effort to displace more and more of the mainstream by engaging with it. Some heterodox economists prefer economics to be thought of as a pluralist discipline, essentially a collection of different approaches to the same topic – see, for example, Fischer et al. (2018). The continuing resistance of the mainstream to criticism is nicely brought out by a recent article analysing the fate of comments on papers in a mainstream economics journal – the *American Economic Review*. Ankel-Peters, Fiala and Neubauer (2025) show that critical comments published in this journal have no impact on the continued citations to the criticized papers.

Earle, Moran and Ward-Perkins's courageous (2017) book, is a detailed look at economics teaching and recruitment in a selection of top UK universities. The book accords with much of what we claim here (but see note 5 in Chapter 6).

7 For a recent historical critique of economics, see Robert Skidelsky's (2020) *What's Wrong with Economics: A Primer for the Perplexed* and his (2018) *Money and Government.* We rely for the cultural influences on economics on Oreskes and Conway's work, which will be discussed at length nearer to the end of the chapter.

8 A graphic illustration of the point that markets give rise to efficiency in a way that planning cannot was provided when one of the authors was walking in Manhattan. A light shower of rain came out of a clear sky and within a couple of minutes umbrellas were on sale at the corner of every block. Planning could not have done it. But the market does not explain why it was not rainhats or, or the sale of magical spells against rain, or just getting wet that was preferred, which might be seen in another society as signifying weakness, or why the umbrella sellers did not prefer to stay at home, sacrificing the money for leisure. Market ideas have a fascinating kernel of truth but explain only a small part of the story.

9 In her book *The World in the Model*, Mary Morgan documents how economics goes from using models to represent some aspect of the world to seeing the world as representing the model and also comments on the implications of this for policy:

> because models operate at a less general level than laws, they tend to embed the normative elements at a level closer to practical matters (however idealized the models themselves might appear to be). Indeed, it is this integration of the normative and positive aspects in models that prompts the way they are taken into the world and used directly as recipes to remake the world, and to change the behaviour of its people, as economists think it and they should function – that is, according to their models.
>
> (Morgan 2012: 404)

10 For negative reviews of the place of string theory in physics, see, for example, Smolin (2008); Hossenfelder (2018).

11 Even adventurous physics such as the detection of gravitational waves is not as dependent on mathematics as many commentators, including some physical scientists, think. The very success of computers has shown that the mathematics at the heart of many such sciences is not the essence of the science since nowadays computers can do most of maths without the need for human brilliance. Even if human mathematicians are needed, they can simply supply their labour – not every physicist needs to be a gifted mathematician any more than every physicist needs to be a gifted experimentalist. See Collins (2007) for an analysis of the role of mathematics in physics.

12 The very idea of diminishing marginal utility, on which the mathematics of markets is based, and which is meant to be the basis of capitalist economies, depends on their having a culture in which the marginal utility of wealth does not diminish. Max Weber's analysis of how a culture based on an interpretation of Protestantism insinuated itself upstream of economics was intended to explain how capitalism came to exist in the first place. Culture is not only upstream of politics it is upstream of economics!

13 Source: BBC iPlayer – The Trap – 1. F**k You Buddy.

14 The consequences of meritocracy taken to its extremes are dramatically portrayed in the South Korean TV series, *Squid Game*, distributed on Netflix.

15 Though Hayek, according to Curtis's documentary: BBC Blogs – Adam Curtis – THE CURSE OF TINA, first advocated the funding of think tanks to bring about political changes in favour of markets. Though *The Road to Serfdom* seems a very reasonable book to a modern reader who has seen Hayek described as the founder of neoliberalism, his later books, such as *The Constitution of Liberty, 1960* (notoriously carried around in Margaret Thatcher's handbag), may be much less reasonable.

16 The contemporary association of evangelical Christianity and free markets is documented by Oreskes as Conway and will be discussed below but it goes back a long way. Exactly how historically accurate this is we do not know, but the way religion and free markets can work together is wonderfully described by historical novelist Amitav Ghosh in his history-based novels about the Opium trade as expressed in the following imagined dialogue between a critic and a major beneficiary of the trade:

> Are you not aware that with every shipment [of opium] you are condemning hundreds, maybe thousands, of people to death? Do you see nothing monstrous in your own actions?
>
> 'No sir' answered Mr Burnham coolly. 'Because it is not my hand that passes sentence on those who choose the indulgence of opium. It is the work of another, invisible, omnipotent: it is the hand of freedom, of the market, of the spirit of liberty itself, which is none other than the breath of God'.
>
> (Ghosh 2011: 486–7)

17 For example, savings rates might be reduced by higher inflation as this typically erodes their value but might also be increased if the rise in inflation is seen as a harbinger of worse times ahead, thus prompting a more precautionary approach.

18 Turchin (2023) is an attempt to create a science of history not subject to historicism.

19 In Chapter 7, we will present the public's failing to support Liz Truss's cutting of the top rate of income tax as a reassuring counter example.

20 Collins and Pinch (1979) shows that journalists occasionally confuse the constitutive and contingent forums. They do this only in the case of low-status sciences. For instance, journalists might launch their own interpretations of experiments involving paranormal research and present their own definitive scientific conclusions (sometimes aided by stage magicians). In ubiquitous economics there is, likewise, no established distinction between the constitutive and contingent forums of economics though the status of the science is higher.
21 Source: https://www.nobelprize.org/prizes/economic-sciences/1974/hayek/speech/.
22 And they might soon do so given the powerful role in government given to Robert F. Kennedy Jnr.
23 Marxist economics and politics could be represented in a similar way, but it has almost no influence in the West these days.
24 Source: Oreskes and Conway (2023: 9).
25 A sketch of von Mises' career can be found at: https://www.econlib.org/library/Enc1/bios/Mises.html.
26 To see why these feedback loops tend to favour the right rather than the left, see Hertel-Fernandez, Skocpol and Sklar (2018).
27 This passage about Fisher and the advice he took from Hayek is taken from Adam Curtis's documentary: BBC Blogs – Adam Curtis – THE CURSE OF TINA. Our main source, however, is Oreskes and Conway (2023), backed up by the various writings of Philip Mirowski.
28 Oreskes and Conway (2023) (tell us that *The Road to Serfdom* was also promulgated in cartoon form in 1945, by the magazine *Look*, and this version was in turn distributed by General Motors(!) as a booklet.
29 For a version of Fisher's encounter with Hayek stressing the contribution of other company donors to the change in economic culture in Britain and the influence of American economists, see Stedman Jones and Jackson (2016).
30 For the story of the evolution of market-based economic ideas, see Mirowski and Plehwe (2009); Mirowski (1999, 2002).
31 One example of how uncertainty and complexity are stripped out of public debate is the way economic forecasts for GDP and other metrics are routinely reported as point estimates rather than ranges. The difference representing the uncertainty would make is clearly illustrated in the Bank of England's *Monetary Policy Report*, in which forecasts for inflation and GDP are presented using 'fan charts' that show how the uncertainty associated with each forecast increases as the forecast period gets longer. For example, the November 2023 report includes a diagram that suggests GDP growth at the end of 2024 (i.e. 12–15 months into the future) could be anywhere between roughly ±4 per cent, with a 10 per cent chance it will lie outside even this range. The same would be true for any other economic forecast;

the only thing that is unusual about the Bank of England is that it is explicit about this.

32 See Collins and Evans (2017b) for a discussion on how to make policy drawing on various different kinds of science with different levels of consensus and drawing on the wisdom of a body of 'Owls' to understand both scientific and social factors.

33 In political theory, the classic distinction is between negative and positive liberty. Negative liberty is freedom from constraints and is the form of freedom that is championed by the proponents of market fundamentalism as they seek to dismantle regulations and other systems that prevent individual choices being enacted. Positive liberty, in contrast, is defined as 'freedom to' and typically requires some kind of regulation or rule to create the 'space' within which this freedom can be exercised.

34 See: https://quoteinvestigator.com/2011/10/15/liberty-fist-nose/ [accessed 9 November 2023].

35 It is interesting that the expression 'nanny state' reflects the socialization practices of the rich: the poor do not have nannies to bring up their children.

36 For an analysis of an American election in these terms, see Collins, Finn and Sutton (2001). We show that the result of the year 2000 election in the United States, taken to have expressed 'the will of the American people', was a matter of statistical noise. See Collins et al. (2020: 2) for a longer analysis of the relationship between the enumerative and organic faces of society. Also note that election winners in both the United States and the United Kingdom are almost always losers in terms of share of the vote because of the way rural voters are given priority over urban voters in the way constituencies are constructed.

37 For example, the UK Bank of England routinely includes fan diagrams that show the uncertainty associated with its main forecasts but these are rarely, if ever, reported in discussions about its interest rate decisions.

38 See Collins and Evans (2017a).

39 A more sophisticated understanding of the sciences reveals that some of the qualitative findings of the social sciences are sounder and more quintessentially testable than those of the quantitative sciences. We have already given one example with the regular place of the verb in the English sentence. Another is given in respect of bus rides in unfamiliar places in Collins (2019: 10–13). On the matter of falsifiability in economics, modelling as a technique that bears on the world is continually being vitiated by the readiness of modellers to adjust assumptions so as to preserve their apparent utility or claim that their models are not wrong but the world is wrong. Furthermore, the mental gymnastics required to maintain that self-sacrifice is utility maximization is a paradigm of unfalsifiability. Though we should make it clear that we are not offering sociology as it is contemporaneously practiced as a fulcrum science either.

40 Skidelsky and Craig (2016: 2) write 'The supposedly neutral discipline of economics does not simply describe human behaviour, but, in fact, shapes it'. This is from their Introduction to their edited book (Skidelsky and Craig 2016b). This book contains a number of discussions of the relationship of economics and political power. Here we are showing one way in which it works. We are saying that neoliberalism leads to the erosion of truth and this can lead to populism and thence to fascism. Mattei (2022) argues that 'austerity', which is said to be necessary to balance the books at times of recession in a country according to mainstream economics (the opposite approach to Keynesianism), was originally intended to cow the working class between the wars and maintain what she calls 'The Capital Order'. According to her analysis this feature of neoliberalism, which is still at play, was a cause of interwar fascism.

Chapter 7

1 At the time of writing, the Conservative Party is choosing a new leader following the crushing defeat in the 2024 election. Astonishingly, after the Truss debacle, they have not changed the selection method: the final choice will be made by a small body of party activists whose predilections are well-understood.

2 An account, though from an activist source, can be found here: https://www.youtube.com/watch?v=IRDLIOME47c.

3 See, for example, https://ourworldindata.org/grapher/tax-revenues-vs-gdp-per-capita.

4 See, for example, the *World Happiness Report* for 2023, available at https://worldhappiness.report/ed/2023/. The rankings can be found in chapter two.

5 At about the same time as she appointed Kwarteng as chancellor, Truss also sacked the most senior civil servant in the Treasury, Tom Scholar. The move was widely seen as another indication of her determination to break with the economic consensus and implement a radical change in economic policy.

6 Source: https://www.telegraph.co.uk/news/2022/09/23/kwasi-kwartengs-budget-moment-history-will-radically-transform/.

7 Who seemed to include among others: 'Labour, the Lib Dems and the SNP, the militant unions, the vested interests dressed up as think tanks, the talking heads, the Brexit deniers and Extinction Rebellion.'

8 And in sympathy with Project 2025's proposal (see below) to replace civil servants with political appointees.

9 To some extent, this representation of Margaret Thatcher as solely interested in tax cuts misrepresents her politics and fails to explain her legacy. For example, in her

early budgets, Thatcher cut direct taxes but simultaneously introduced big increases to VAT. More important, however, is the idea that changes to the tax system were not an end in themselves but a part of an overall strategy to fundamentally change the culture and values of British society. As she herself put it: 'Economics is the method: the object is to change the heart and soul' (https://www.margaretthatcher.org/document/104475).

10 Thanks to Paul Whiteley and Josh Ryan-Collins for advice on this paragraph. For a recent analysis, that finds no evidence to support the idea that tax cuts for the wealthy either reduce inequality or promote GDP growth, see Hope and Limberg (2022).

11 Source: https://www.itv.com/news/2022-09-20/biden-slams-trickle-down-economics-as-truss-plans-to-lift-bankers-bonus-cap.

12 There is no low-growth coalition as claimed by Truss: no potential UK government is against growth, so it would be widely agreed that the UK cannot afford to be a low-growth economy. The question is whether low growth follows from a larger government, for which there is no evidence, at least up to a point.

13 See, for example, https://www.oecd.org/tax/tax-policy/revenue-statistics-highlights-brochure.pdf.

14 The United States seems different in this respect, large parts of the population being ready to continue to endorse a presidential candidate who has no interest in the truth.

15 The tense of much of this paragraph has had to be changed from present to past!

16 As had been the overwhelming defeat of the immensely experienced but vocal anti-Trump incumbent, Liz Cheney, in the August 2022 Republican Primary for the House of Representatives. Of course, the remaining positive indicators in the United States have now been nullified.

17 A similar argument can be made in respect of the 2023 general elections in Poland, where the populist Law and Justice Party (known as PiS) was voted out of office after eight years in power. Although still the largest party, the PiS was unable to form the coalition needed to retain office. Instead, the Civic Coalition, led by former President of the European Council Donald Tusk, formed the new government. Significantly, the election had a turnout of nearly 73 per cent, suggesting that many citizens knew exactly what was at stake. Information about election results from: https://www.bbc.com/news/world-europe-67118787.

18 For a first discussion of this law, see Collins and Evans (2017b).

19 With the results of the recent Polish election (note 17 in Chapter 7) and the fall of Bolsanaro in Brazil, also suggesting that experiences of populist government can remind people of why the alternative matters.

20 And see the claim in Levitsky and Ziblatt (2023) that a material/cultural cause of Trumps success in the United States is the fear of the demise of the country as a white Christian nation.

21 Brown (2019) describes, in her introductory chapter, the many changes in Western societies that have led in this direction.

22 See, for example, van Leent et al. (2023); WHO (2018).

23 Unfortunately, research students in the humanities and social science in the United States are in a difficult position. In these domains, the route to a PhD is a difficult one and the route to a tenured position at a university is still more difficult. American PhD and post-doc students in these subjects are, therefore, in thrall to their supervisors and colleagues. This makes academic change in the humanities and social sciences, which is really social change, very slow. Turning the key in respect of the old positions on MMR would help a lot.

24 Farmers and farmworkers should sometimes be included but not MMR or COVID anti-vaxxers; the democratizers have to explain the difference if their programme is to have integrity.

25 The lecturer was Professor Ben Ansell from Oxford University though his concerns seem a long way from ours.

26 Turchin (2023) is a counter-example.

27 For example, the 2023 Dutch election in which the populist Geert Wilders' Party for Freedom (PVV) emerged as the leader of the largest party. The concern is that this is due, at least in part, to the way that mainstream parties accommodate and normalize their concerns by 'agreeing' to campaign on the same issues.

28 As it was not in the last US election. For a right-wing rant favouring market forces, see Jordan Peterson in *The Telegraph* (note 10 in Chapter 2).

29 Batkie (2023, 2021) surveys American science teachers on the way they see their responsibility towards civic education and finds that they believe teaching 'business as usual' science facts is what is needed. This attitude will need to change if democracy is to be saved in the schoolroom by teaching students to understand how science works. Collins, Batke and Zodda (an American high school teacher) are hoping to publish a special section in a science education-related journal putting forward a new syllabus and describing its problems.

30 Unfortunately, the events of 2024 show that they now constitute a minority of US voters.

31 There is a parallel with the United States in the Conservative Party in that the post-resignation party leader is chosen by a relatively small number of Conservative activists, not the parliamentary party or the people as a whole, who must await another election to have their say. Conservative Party activists are untypical of general opinion. This is a bad way to run a party or a country as the Truss episode dramatically illustrates.

32 Declaration made on 8 July in response to Johnson's 'Partygate' scandal and his brushing-off of a police fine, with Starmer's own promise to step down if found

guilty over the supposedly equivalent 'Beergate' claims – he was subsequently cleared of wrongdoing. Sadly, Starmer let himself down (early April 2023) by endorsing a Labour Party advertising campaign claiming that Sunak did not believe paedophiles should go to prison. Reassuringly, this caused a huge backlash at all levels in the Labour Party.

33 For an analysis of the role of social media, see Collins et al. (2022) and for the potential of Generative AI, see Collins (2024).

References cited

Abbot, D., A. Bikfalvi, A. L. Bleske-Rechek, W. Bodmer, P. Boghossian, C. M. Carvalho, J. Ciccolini, J. A. Coyne, J. Gauss, P. M. W. Gill, S. Jitomirskaya, L. Jussim, A. I. Krylov, G. C. Loury, L. Maroja, J. H. McWhorter, S. Moosavi, P. Nayna Schwerdtle, J. Pearl, M. A. Quintanilla-Tornel, H. F. Schaefer Iii, P. R. Schreiner, P. Schwerdtfeger, D. Shechtman, M. Shifman, J. Tanzman, B. L. Trout, A. Warshel, and J. D. West. 2023. 'In Defense of Merit in Science'. *Controversial Ideas* 3(1): 0–0. doi: 10.35995/jci03010001.

Ankel-Peters, Jörg, Nathan Fiala, and Florian Neubauer. 2025. 'Is Economics Self-Correcting? Replications in the *American Economic Review*'. *Economic Inquiry* 63(2): 463–85. https://doi-org.cardiff.idm.oclc.org/10.1111/ecin.13222.

Arendt, Hannah. 1951. *The Origins of Totalitarianism*. New York: Harcourt, Brace.

Arksey, Hilary. 1998. *RSI and the Experts: The Construction of Medical Knowledge*. London; Bristol, PA: UCL Press.

Azam, Jean-Paul. 2005. 'Suicide-Bombing as Inter-Generational Investment'. *Public Choice* 722(1–2): 177–98.

Balmer, Brian. 2012. *Secrecy and Science: A Historical Sociology of Biological and Chemical Warfare*. London: Routledge.

Barnes, Barry, David Bloor, and John Henry. 1996. *Scientific Knowledge: A Sociological Analysis*. Chicago, IL: University of Chicago Press.

Batkie, Ryan. 2023. 'Secondary Science Teacher Beliefs about Science Education for Citizenship'. PhD thesis in Curriculum and Instruction, University of Wisconsin-Madison.

Batkie, Ryan. 2021. 'Considering Science Education's Civic Responsibility: Mapping a Dilemma for the Field'. *American Educational Research Association*. doi: https://doi.org/10.3102/1691273.

Berman, Eli and David D. Laitin 2008. 'Religion, Terrorism and Public Goods: Testing the Club Model'. *Journal of Public Economics* 92(10–11): 1942–67.

Bloor, David 1973. 'Wittgenstein and Mannheim on the Sociology of Mathematics'. *Studies in the History and Philosophy of Science* 4(2): 173–91.

Bloor, David. 1991. *Knowledge and Social Imagery*. 2nd ed. Chicago, IL: University of Chicago Press.

Brown, Wendy. 2019. *In the Ruins of Neoliberalism: The Rise of AntiDemocratic Politics in the West*. New York: Columbia University Press.

Callard, Felicity and Elisa Perego. 2021. 'How and Why Patients *Made* Long Covid'. *Social Science & Medicine* 268: 113426. doi: 10.1016/j.socscimed.2020.113426.
Campo-Engelstein, Lisa and Nadia L. Johnson. 2014. 'Revisiting "The Fertilization Fairytale": An Analysis of Gendered Language Used to Describe Fertilization in Science Textbooks from Middle School to Medical School'. *Cultural Studies of Science Education* 9(1): 201–20. doi: 10.1007/s11422-013-9494-7.
Cartwright, Nancy and John Bryan Davis. 2016. 'Economics as Science'. pp. 43–55 in *Who Runs the Economy? The Role of Power in Economics*, edited by R. Skidelsky and N. Craig. London: Palgrave Macmillan UK.
Caudill, David S., Harry Collins, and Robert Evans. 2024. 'Judges Should Be Discerning Consensus, Not Evaluating Scientific Expertise'. *University of Cincinnati Law Review* 92(4): 1031–79. Available from: https://scholarship.law.uc.edu/uclr/vol92/iss4/11/.
Caudill, David S., Shannon N. Conley, Michael E. Gorman, and Martin Weinel, eds. 2019. *The Third Wave in Science and Technology Studies: Future Research Directions on Expertise and Experience*. Cham, Switzerland: Palgrave MacMillan.
Collins, Harry. 1974. 'The TEA Set: Tacit Knowledge and Scientific Networks'. *Science Studies* 4(2): 165–86.
Collins, Harry. 1975. 'The Seven Sexes: A Study in the Sociology of a Phenomenon, or The Replication of Experiments in Physics'. *Sociology* 9(2): 205–24.
Collins, Harry. 1981. 'Introduction: Stages in the Empirical Programme of Relativism'. *Social Studies of Science* 11(1): 3–10.
Collins, Harry. 1985/1992. *Changing Order: Replication and Induction in Scientific Practice*. Beverley Hills, CA, & London: Sage [2nd ed. 1992, Chicago, IL: University of Chicago Press].
Collins, Harry. 1996. 'In Praise of Futile Gestures: How Scientific Is the Sociology of Scientific Knowledge'. *Social Studies of Science* 26(2): 229–44.
Collins, Harry. 2001. 'Tacit Knowledge, Trust, and the Q of Sapphire'. *Social Studies of Science* 31(1): 71–85.
Collins, Harry. 2004a. 'Interactional Expertise as a Third Kind of Knowledge'. *Phenomenology and the Cognitive Sciences* 3(2): 125–43.
Collins, Harry. 2004b. Gravity's *Shadow: The Search for Gravitational Waves*. Chicago, IL: The University of Chicago Press.
Collins, Harry. 2007. 'Mathematical Understanding and the Physical Sciences'. Harry Collins (ed) *Case Studies of Expertise and Experience: Special Issue of Studies in History and Philosophy of Science* 38(4) (December): 667–85.
Collins, Harry. 2008. 'The Meaning of Hoaxes'. pp. 77–81 in *Knowledge as Social Order: Rethinking the Sociology of Barry Barnes*, edited by Massimo Mazzotti. London: Ashgate.
Collins, Harry. 2011. 'Language and Practice'. *Social Studies of Science* 41(2): 271–300. doi: 10.1177/0306312711399665.

Collins, Harry. 2013. *Gravity's Ghost and Big Dog: Scientific Discovery and Social Analysis in the Twenty-First Century*. Enlarged ed. Chicago, IL: The University of Chicago Press.

Collins, Harry. 2014. 'Rejecting Knowledge Claims Inside and Outside Science'. *Social Studies of Science* 44(5): 722–35.

Collins, Harry. 2017. *Gravity's Kiss: The Detection of Gravitational Waves*. Cambridge, MA: MIT Press.

Collins, Harry. 2018. *Artifictional Intelligence: Against Humanity's Surrender to Computers*. Cambridge: Polity Press.

Collins, Harry. 2019. *Forms of Life: The Method and Meaning of Sociology*. Cambridge, MA: MIT Press.

Collins, Harry. 2020. 'Interactional Imogen: Language, Practice and the Body'. *Phenomenology and the Cognitive Sciences*. doi: 10.1007/s11097-020-09679-x.

Collins, Harry. 2024. 'Why Artificial Intelligence Needs Sociology of Knowledge: Parts I and II'. *AI and Society*. https://doi.org/10.1007/s00146-024-01954-8.

Collins, Harry and Martin Kusch. 1998. *The Shape of Actions: What Humans and Machines Can Do*. Cambridge, MA: MIT Press.

Collins, Harry and Robert Evans. 2002. 'The Third Wave of Science Studies: Studies of Expertise and Experience'. *Social Studies of Science* 32(2): 235–96. doi: 10.1177/0306312702032002003.

Collins, Harry and Robert Evans. 2007. *Rethinking Expertise*. Chicago, IL: University of Chicago Press.

Collins, Harry and Robert Evans. 2015. 'Expertise Revisited, Part I – Interactional Expertise'. *Studies in History and Philosophy of Science Part A* 54: 113–23. doi: 10.1016/j.shpsa.2015.07.004.

Collins, Harry and Robert Evans. 2017a. 'Probes, Surveys and the Ontology of the Social'. *Journal of Mixed Methods Research* 11(3): 328–42 (doi/abs/10.1177/1558689815619825).

Collins, Harry and Robert Evans. 2017b. *Why Democracies Need Science*. Cambridge: Polity Press.

Collins, Harry and Trevor Pinch. 1979. 'The Construction of the Paranormal: Nothing Unscientific Is Happening'. *The Sociological Review* 27(1_suppl): 237–70. doi: 10.1111/j.1467-954X.1979.tb00064.x.

Collins, Harry and Trevor Pinch. 1993/1998. *The Golem: What Everyone Should Know about Science*. Cambridge & New York: Cambridge University Press [New ed. 1998].

Collins, Harry, Andrew Bartlett, and Luis Reyes-Galindo. 2017. 'Demarcating Fringe Science for Policy'. *Perspectives on Science* 25(4): 411–38. doi: 10.1162/POSC_a_00248.

Collins, Harry, Lee Samuel Finn, and Patrick Sutton. 2001. 'What Is TWAP?: Three Comments on the American Election of the Year 2000'. *Social Studies of Science* 31(3): 428–36.

Collins, Harry, Robert Evans, and Luis Reyes Galindo. 2024. 'Virtual Diversity: Resolving the Tension between the Wider Culture and the Institution of Science'. *Social Studies of Science*. Available online at doi: 10.1177/03063127241263609.

Collins, Harry, Robert Evans, and Luis Reyes Galindo. 2025. 'Virtual Diversity Revisited'. *Social Studies of Science* 55(2): 316–24.

Collins, Harry, Robert Evans, and Martin Weinel. 2017. 'Interactional Expertise'. pp. 765–92 in *Handbook of Science and Technology Studies*, edited by C. A. Miller, L. Smith-Doerr, U. Felt, and R. Fouche. Cambridge, MA: MIT Press.

Collins, Harry, Robert Evans, and Michael Gorman. 2007. 'Trading Zones and Interactional Expertise'. Collins (ed) *Case Studies of Expertise and Experience: special issue of Studies in History and Philosophy of Science* 38(4): 657–66.

Collins, Harry, Robert Evans, and Michael Gorman. 2017/2019. 'Trading Zones Revisited'. http://arxiv.org/abs/1712.06327 and pp. 275–81 in *The Third Wave in Science and Technology Studies: Future Research Directions on Expertise and Experience*, edited by David Caudill et al. Cham, Switzerland: Palgrave, MacMillan.

Collins, Harry, Willow Leonard-Clarke, and Will Mason-Wilkes. 2022. 'Scientific Conferences, Socialisation and Lockdown'. *Social Studies of Science* 53(3): 379–401. doi: 10.1177/03063127221138521.

Collins, Harry, Robert Evans, Darrin Durant, and Martin Weinel. 2020. *Experts and the Will of the People: Society, Populism and Science*. Cham, Switzerland: Palgrave Pivot.

Collins, Harry, Jeff Shrager, Andy Bartlett, Shannon Conley, Rachel Hale, and Robert Evans. 2023. 'Hyper-Normal Science and Its Significance'. *Perspectives on Science* 31(2): 262–92. doi: 10.1162/posc_a_00572.

Collins, Harry, Robert Evans, Martin Innes, Eric B. Kennedy Kennedy, Will Mason-Wilkes, and John McLevy. 2022. *The Face-to-Face Principle: Science, Trust, Democracy and the Internet*. Cardiff: Cardiff University Press Open Access. https://doi.org/10.18573/book7.

Dawkins, Richard. 1976. *The Selfish Gene*. New York: Oxford University Press.

Dawkins, Richard. 1986. *The Blind Watchmaker: Why the Evidence of Evolution Reveals a Universe without Design*. New York: Norton.

Deneen, Patrick J., Francis Fukyama, Deirdre N. McCloskey, and West Cornell. 2023. 'Is Liberalism Worth Saving'. *Harper's Magazine* 23–26, Forum, edited by Christopher Beha, February.

Dewey, John. 1927/1954. *The Public and Its Problems*. Athens, OH: Swallow Press (Original work published 1927.)

Douglas, Heather. 2009. *Science, Policy and the Value-Free Ideal*. Pittsburgh, PA: University of Pittsburgh Press.

Douglas, Heather. 2021. *The Rightful Place of Science: Science, Values and Democracy*. Tempe, AZ: Consortium for Science Policy and Outcomes.

Du Bois, W. E. B. 1903/1994. *The Souls of Black Folk*. Avenel: Gramercy Books.

Duhem, Pierre. 1908/1981. *The Aim and Structure of Physical Theory*. New York: Athenaeum 1981 [First published in 1908].
Durkheim, Emile. 1958. *Professional Ethics and Civic Morals*. Translated by Cornelia Brookfield. Glencoe, IL: Free Press.
Durkheim, Emile. 2019. *Professional Ethics and Civic Morals*. With an Introduction by Bryan Turner. New York: Routledge Classics.
Earle, Joe, Cahal Moran, and Zach Ward-Perkins. 2017. *The Econocracy: On the Perils of Leaving Economics to the Experts*. Manchester: Manchester University Press.
Eddo-Lodge, Reni. 2017. 'Why I'm No Longer Talking to White People about Race'. *The Guardian*, 30 May. Available online at: https://www.theguardian.com/world/2017/may/30/why-im-no-longer-talking-towhite-people-about-race.
Eddo-Lodge, Reni. 2020. *Why I'm No Longer Talking to White People about Race*. London: Bloomsbury.
Elliott, Kevin C. 2022. *Values in Science*. Cambridge: Cambridge University Press.
Epstein, Steven. 1995. 'The Construction of Lay Expertise: AIDS Activism and the Forging of Credibility in the Reform of Clinical Trials'. *Science, Technology, & Human Values* 20(4): 408–37. doi: 10.1177/016224399502000402.
Epstein, Steven. 1996. *Impure Science: AIDS, Activism, and the Politics of Knowledge*. Berkeley: University of California Press.
Evans, Robert. 2007. 'Social Networks and Private Spaces in Economic Forecasting'. *Studies in History and Philosophy of Science Part A* 38(4): 686–97. doi: 10.1016/j.shpsa.2007.09.011.
Eyal, Gil. 2013. 'For a Sociology of Expertise: The Social Origins of the Autism Epidemic'. *American Journal of Sociology* 118(4): 863–907. doi: 10.1086/668448.
Eyal, Gil. 2019. *The Crisis of Expertise*. London: Polity.
Fischer, Liliann, Joe Hasell, J. Christopher Proctor, David Uwakwe, Zach Ward-Perkins, and Catriona Watson, eds. 2018. *Rethinking Economics: An Introduction to Pluralist Economics*. Abingdon, OX; New York: Routledge.
Fleck, Ludwik. 1935/1979. *Genesis and Development of a Scientific Fact*. Chicago: University of Chicago Press (first published in German in 1935 as *Entstehung und Entwicklung einer wissenschaftlichen Tatsache: Einführung in die Lehre vom Denkstil und Denkkollektiv*).
Fourcade, Marion. 2006. 'The Construction of a Global Profession: The Transnationalization of Economics'. *American Journal of Sociology* 112(1): 145–94. doi: 10.1086/502693.
Freedland, Jonathan. 2022. 'The Reality of Brexit Is Biting Hard – Now Everyone Can See It'. *Guardian, The Journal* P1, 3 December.
Friedman, Milton. 1962. *Capitalism and Freedom*. Chicago, IL: The University of Chicago Press.
Funtowicz, Silvio O. and Jerome R. Ravetz. 1993. 'Science for the Post-Normal Age'. *Futures* 25(7): 739–55. doi: 001632879390022L.

Galison, Peter Louis. 1997. *Image and Logic a Material Culture of Microphysics*. Chicago, IL: University of Chicago Press.

Ghosh, Amitav. 2011. *River of Smoke*. London: John Murray.

Goodman, Nelson. 1955/1973. *Fact, Fiction, and Forecast*. 3rd ed. Indianapolis, IN: Bobbs-Merrill.

Gorman, Michael E., ed. 2010. *Trading Zones and Interactional Expertise Creating New Kinds of Collaboration*. Cambridge, MA: MIT Press.

Harari, Yuval Noaḥ. 2015. *Sapiens: A Brief History of Humankind*. First U.S. ed. New York: HarperCollins Publishers.

Harding, Sandra. 1995a. 'Can Feminist Thought Make Economics More Objective?' *Feminist Economics* 1(1): 7–32. doi: 10.1080/714042212.

Harding, Sandra. 1995b. '"Strong Objectivity": A Response to the New Objectivity Question'. *Synthese* 104(3): 331–49. doi: 10.1007/BF01064504.

Hawking, Stephen. 1988. *A Brief History of Time*. London: Bantam Dell.

Hayek, Friedrich. 1944. *The Road to Serfdom*. London: Routledge.

Hayek, Friedrich. 1960. *The Constitution of Liberty*. Chicago, IL: University of Chicago Press.

Hertel-Fernandez, Alex, Theda Skocpol, and Jason Sklar. 2018. 'When Political Mega-Donors Join Forces: How the Koch Network and Democracy Alliance Influence Organized US Politics'. *Studies in American Political Development* 2(32): 127–16. doi: 10.1017/S0898588X18000081.

Hollinger, David A. 1996. 'The Defense of Democracy and Robert K. Merton's Formulation of the Scientific Ethos'. pp. 80–96 in *Science, Jews, and Secular Culture: Studies in Mid-Twentieth-Century American Intellectual History*, edited by David A. Hollinger. Princeton, NJ: Princeton University Press.

Hope, David and Julian Limberg. 2022. 'The Economic Consequences of Major Tax Cuts for the Rich'. *Socio-Economic Review* 20(2): 539–59. doi: 10.1093/ser/mwab061.

Hossenfelder, Sabine. 2018. *Lost in Math*. New York: Basic Books.

Jasanoff, Sheila. 2003. Breaking the Waves in Science Studies: Comment on H. M. Collins and Robert Evans, 'The Third Wave of Science Studies'. *Social Studies of Science* 33(3): 389–400.

Jasanoff, Sheila, ed. 2004. *States of Knowledge: The Co-Production of Science and Social Order*. London; New York: Routledge.

Jasanoff, Sheila. 2013. 'Fields and Fallows: A Political History of STS'. pp. 99–118 in *Interdisciplinarity: Reconfigurations of the Natural and Social Sciences*, edited by A. Barry and G. Born. London; New York: Routledge.

Kakutani, Michiko. 2018. *The Death of Truth*. New York: Tim Duggan Books.

Krige, John. 2001. 'Distrust and Discovery: The Case of the Heavy Bosons at CERN'. *Isis* 92(3): 517–40.

Krukones, Michael G. 1985. 'The Campaign Promises of Jimmy Carter: Accomplishments and Failures'. *Presidential Studies Quarterly* 15(1): 136–44.
Kuhn, Thomas S. 1959. 'The Essential Tension: Tradition and Innovation in Scientific Research., Alta, 11–14 June 1959 (pp. 162–174). Salt Lake City: University of Utah Press (Conference in Alta, 11–14 June 1959)'. pp. 162–74 in *The Third University of Utah Conference on the Identification of Creative Scientific Talent (Confernece in Alat, 11–14 June 1959)*, edited by C. W. Taylor. Salt Lake City: University of Utah Press.
Kuhn, Thomas S. 1962. *The Structure of Scientific Revolutions*. Chicago, IL: University of Chicago Press.
Kuhn, Thomas S. 1977. *The Essential Tension: Selected Studies in Scientific Tradition and Change*. Chicago, IL: University of Chicago Press.
Latour, Bruno. 1993. *We Have Never Been Modern*. Cambridge, MA: Harvard University Press.
Latour, Bruno. 2005. *Reassembling the Social: An Introduction to Actor-Network-Theory*. Oxford; New York: Oxford University Press.
Latour, Bruno and Steve Woolgar. 1979. *Laboratory Life: The Social Construction of Scientific Facts*. Beverly Hills, CA: Sage Publications.
van Leent, Lisa, Kerryann Walsh, Kirstine Hand Claire Moran, and Sally French. 2023. 'Effectiveness of Relationships and Sex Education: A Systematic Review of Terminology, Content, Pedagogy, and Outcomes'. *Educational Research Review* 39: 100527. doi: 10.1016/j.edurev.2023.100527.
Levitsky, Steven and Daniel Ziblatt. 2018. *How Democracies Die*. 1st ed. New York: Crown.
Levitsky, Steven and Daniel Ziblatt. 2023. *Tyranny of the Minority: How to Reverse an Authoritarian Turn, and Forge a Democracy for All*. London: Viking.
Lewis, Michael. 1989. *Liar's Poker*. London: Hodder and Staughton.
Lippmann, Walter. 1925. *The Phantom Public*. New York: Harcourt, Brace and Company.
Longino, Helen E. 1990. *Science as Social Knowledge: Values and Objectivity in Scientific Inquiry*. Princeton, NJ: Princeton University Press.
Longino, Helen E. 2002. *The Fate of Knowledge*. Princeton, NJ: Princeton University Press.
Ludwig, David and Luana Poliseli. 2018. 'Relating Traditional and Academic Ecological Knowledge: Mechanistic and Holistic Epistemologies across Cultures'. *Biology & Philosophy* 33(5): 43. doi: 10.1007/s10539-018-9655-x.
MacKenzie, Donald A. 1981. *Statistics in Britain, 1865–1930*. Edinburgh: Edinburgh University Press.
Macpherson, William. 1999. *The Stephen Lawrence Inquiry*. Accessed 13 September 2024. Cm-4262-I. Available from: https://www.gov.uk/government/publications/the-stephen-lawrence-inquiry.

Martin, Emily. 1991. 'The Egg and the Sperm: How Science Has Constructed a Romance Based on Stereotypical Male-Female Roles'. *Signs* 16(3): 485–501.
Mattei, Clara E. 2022. *The Capital Order: How Economists Invented Austerity and Pave the Way to Fascism*. Chicago, IL: University of Chicago Press.
McCright, Aaron M. 2016. 'Anti-Reflexivity and Climate Change Skepticism in the US General Public'. *Human Ecology Review* 22(2): 77–108. http://www.jstor.org/stable/24875159.
Merton, Robert K. 1942. 'Science and Technology in a Democratic Order'. *Journal of Legal and Political Sociology* (1 and 2) (October): 115–26.
Mirowski, Philip. 1999. *More Heat than Light: Economics as Social Physics: Physics as Nature's Economics*. 1st paperback ed., reprinted. Cambridge: Cambridge University Press.
Mirowski, Philip. 2002. *Machine Dreams: Economics Becomes a Cyborg Science*. Cambridge; New York: Cambridge University Press.
Mirowski, Philip. 2020. 'Democracy, Expertise and the Post-Truth Era: An Inquiry into the Contemporary Politics of STS'. Version 1.1, April 2020.
Mirowski, Philip. Forthcoming. *Open Science and Its Frenemies*. Cambridge, MA: MIT Press.
Mirowski, Philip and Dieter Plehwe, eds. 2009. *The Road from Mont Pèlerin: The Making of the Neoliberal Thought Collective*. Cambridge, MA: Harvard University Press.
Monbiot, George and Peter Hutchison. 2024. *The Invisible Doctrine: The Secret History of Neoliberalism (& How It Came to Control Your Life)*. London: Penguin.
Morelock, Jeremiah and Felipe Ziotti Narita. 2022. 'The Nexus of QAnon and COVID-19: Legitimation Crisis and Epistemic Crisis'. *Critical Sociology* 48(6): 1005–24. https://doi.org/10.1177/08969205211069614.
Morgan, Mary S. 2012. *The World in the Model: How Economists Work and Think*. Cambridge: Cambridge University Press.
Mounk, Yascha. 2018. 'How Populist Uprisings Could Bring down Liberal Democracy | Yascha Mounk'. *The Guardian*, 4 March.
Olson, Mancur. 1965. *The Logic of Collective Action: Public Goods and the Theory of Groups*. Cambridge, MA: Harvard University Press.
Oreskes, Naomi and Erik Conway. 2011. *Merchants of Doubt: How a Handful of Scientists Obscured the Truth on Issues from Tobacco Smoke to Global Warming*. New York: Bloomsbury.
Oreskes, Naomi and Erik Conway. 2023. *The Big Myth: How American Business Taught Us to Loathe Government and Love the Free Market*. New York: Bloomsbury.
Papada, Evie, David Altman, Fabio Angiolillo, Lisa Gastaldi, Tamara Köhler, Martin Lundstedt, Natalia Natsika, Marina Jirotka, Yuko Sato, Felix Wiebrecht, and Staffan I. Lindberg. 2023. *Defiance in the Face of Autocratization: Democracy Report 2023*. Gothenburg: University of Gothenburg: Varieties of Democracy Institute (V-Dem Institute).

Pinch, Trevor, Harry M. Collins, and Larry Carbone. 1996. 'Inside Knowledge: Second Order Measures of Skill'. *The Sociological Review* 44(2): 163–86. doi: 10.1111/j.1467-954X.1996.tb00420.x.

Pomerantsev, Peter. 2014. 'The Hidden Author of Putinism'. *The Atlantic*. Retrieved 19 January 2024 (https://www.theatlantic.com/international/archive/2014/11/hidden-author-putinism-russia-vladislav-surkov/382489/).

Pomerantsev, Peter. 2019. This *Is Not Propaganda: Adventures in the War against Reality*. 1st ed. New York: PublicAffairs.

Popper, Karl R. 1959. *The Logic of Scientific Discovery*. New York: Harper & Row.

Power, Sam. 2023. 'Nadhim Zahawi Sacked: Today's Tory Scandals Are Similar to 1990s Sleaze Stories in More than One Way'. *The Conversation*. Retrieved 18 January 2024 (http://theconversation.com/nadhim-zahawi-sacked-todays-tory-scandals-are-similar-to-1990s-sleaze-stories-in-more-than-one-way-198462).

Rabeharisoa, Vololona and Michel Callon. 2004. 'Patients and Scientists in French Muscular Dystrophy Research'. pp. 142–60 in *States of Knowledge: The Co-Production of Science and Social Order*, edited by S. Jasanoff. London: Routledge.

Renck, Vitor, David Ludwig, Paride Bollettin, and Charbel N. El-Hani. 2022. 'Exploring Partial Overlaps between Knowledge Systems in a Brazilian Fishing Community'. *Human Ecology* 50(4): 633–49. doi: 10.1007/s10745-022-00335-3.

Reyes-Galindo, Luis. 2017. 'Molecular Detector (Non)Technology in Mexico'. *Science Technology & Human Values* 42(1): 86–115.

Reyes-Galindo, Luis. 2023. 'Values and Vendettas: Populist Science Governance in Mexico'. *Social Studies of Science* 53(2): 213–41. doi: 10.1177/03063127221140020.

Rip, Arie. 2003. 'Constructing Expertise in a Third Wave of Science Studies?' *Social Studies of Science* 33(3): 419–34.

Rosser, Sue V. 1992. 'Are There Feminist Methodologies Appropriate for the Natural Sciences and Do They Make a Difference?' *Women's Studies International Forum* 15(5): 535–50. doi: 10.1016/0277-5395(92)90057-3.

Rovelli, Carlo. 2021. *Helgoland*. Dublin: Allen Lane.

Roxby, Philippa. 2024. 'Alarming 45-Fold Rise in Measles - WHO'. *BBC News*, 23 January 2024. Available at: https://www.bbc.co.uk/news/health-68068226.

Scheffler, Israel. 1967. *Science and Subjectivity*. Cambridge, MA: Hackett.

Schudson, Michael. 2008. 'The "Lippmann–Dewey Debate" and the Invention of Walter Lippmann as an Anti-Democrat 1986–1996'. *International Journal of Communication* 2(1): 1031–42.

Scott, Pam, Evelleen Richards, and Brian Martin. 1990. 'Captives of Controversy: The Myth of the Neutral Social Researcher in Contemporary Scientific Controversies'. *Science, Technology, & Human Values* 15(4): 474–94.

Shapin, Steven. 2016. 'Invisible Science'. *The Hedgehog Review* 18(3) (Fall): 34–46. https://scholar.harvard.edu/files/shapin/files/invisible_science_final.pdf.

Shimuzo, Hirofumi. 2011. 'Social Cohesion and Self-Sacrificing Behavior'. *Public Choice* 149: 427–40. doi: 10.1007/sl1127-011-9880-1.

Skidelsky, Robert. 2018. *Money and Government: A Challenge to Mainstream Economics.* New Dehli, India: Allen Lane.

Skidelsky, Robert. 2020. *What's Wrong with Economics: A Primer for the Perplexed.* New Haven, CT: Yale.

Smolin, Lee. 2008. *The Trouble with Physics: The Rise of String Theory, The Fall of a Science and What Comes Next.* London: Penguin Books.

Skidelsky, Robert and Nan Craig. 2016a. 'Introduction'. pp. 1–14 in *Who Runs the Economy: The Role of Power in Economics,* edited by Robert Skidelsky and Nan Craig. Basingstoke: Palgrave Macmillan.

Skidelsky, Robert and Nan Craig, eds. 2016b. *Who Runs the Economy: The Role of Power in Economics.* Basingstoke Palgrave Macmillan.

Stedman Jones, Daniel and Ben Jackson. 2016. 'The Neoclassical Counter-Revolution and the Ascendancy of Business'. pp. 73–84 in *Who Runs the Economy: The Role of Power in Economics,* edited by Robert Skidelsky and Nan Craig. Basingstoke: Palgrave Macmillan.

Steel, Daniel. 2021. 'What about Trust: Communication and Public Controversies about Science, Commentary on Lecture 3'. pp. 167–77 in *The Rightful Place of Science: Science, Values, and Democracy,* edited by H. Douglas. Arizona State University: Center for Science & Policy Outcomes.

Turchin, Peter. 2023. *End Times: Elites, Counter-Elites and the Path of Political Disintegration.* New York: Penguin Books.

Turner, Stephen and Daryl Chubin. 2020. 'The Changing Temptations of Science'. *Issues in Science and Technology* 36(3) (Spring): 40–6.

Wakefield, A. J., S. H. Murch, A. Anthony, J. Linnell, D. M. Casson, M. Malik, M. Berelowitz, A. P. Dhillon, M. A. Thomson, P. Harvey, A. Valentine, S. E. Davies, and J. A. Walker-Smith. 1998. 'RETRACTED: Ileal-Lymphoid-Nodular Hyperplasia, Non-Specific Colitis, and Pervasive Developmental Disorder in Children'. *The Lancet* 351(9103): 637–41. doi: 10.1016/S0140-6736(97)11096-0.

Walker, Rafael. 2022. 'Who Gets to Write about Whom?' *The Chronicle of Higher Education.* Retrieved 15 August 2023 (https://www.chronicle.com/article/who-gets-to-write-about-whom).

Weinel, Martin. 2007. 'Primary Source Knowledge and Technical Decision-Making: Mbeki and the AZT Debate'. *Studies in History and Philosophy of Science Part A* 38(4): 748–60. doi: 10.1016/j.shpsa.2007.09.010.

Weinel, Martin. 2010. 'Technological Decision-Making under Scientific Uncertainty: Preventing Mother-to-Child Transmission of HIV in South Africa'. PhD, Cardiff University.

WHO. 2018. *International Technical Guidance on Sexuality Education: An Evidence-Informed Approach*. Rev. ed. Paris, New York, Geneva: UNESCO : UN-Women : UNICEF : UNFPA : UNAIDS : WHO.

Wickson, Fern and Brian Wynne. 2012. 'The Anglerfish Deception'. *EMBO Reports* 13(2): 100–5. doi: 10.1038/embor.2011.254.

Winch, Peter G. 1958. *The Idea of a Social Science*. London: Routledge and Kegan Paul.

Wittgenstein, Ludwig. 1953. *Philosophical Investigations*. Oxford: Blackwell.

Wynne, Brian. 1989. 'Sheep Farming after Chernobyl: A Case Study in Communicating Scientific Information'. *Environmental Magazine* 31(2): 33–9.

Wynne, Brian. 1992. 'Misunderstood Misunderstanding: Social Identities and Public Uptake of Science'. *Public Understanding of Science* 1(3): 281–304. doi: 10.1088/0963-6625/1/3/004.

Index

www.ingramcontent.com/pod-product-compliance
Ingram Content Group UK Ltd.
Pitfield, Milton Keynes, MK11 3LW, UK
UKHW020212290926
15470UKWH00001B/136

9 781666 980905